Good Science, Bad Science, Pseudoscience, and Just Plain Bunk

Good Science, Bad Science, Pseudoscience, and Just Plain Bunk

How to Tell the Difference

Peter A. Daempfle

ROWMAN & LITTLEFIELD PUBLISHERS, INC.
Lanham • Boulder • New York • Toronto • Plymouth, UK

"Wisdom of Dialectical Materialism" from *Albert Einstein: Creator and Rebel* by Banesh Hoffmann and Helen Dukas, copyright 1972 by Helen Dukas and Banesh Hoffmann. Used by permission of Dutton, a division of Penguin Group (USA) Inc.

Published by Rowman & Littlefield Publishers, Inc.
A wholly owned subsidiary of The Rowman & Littlefield Publishing Group, Inc.
4501 Forbes Boulevard, Suite 200, Lanham, Maryland 20706
www.rowman.com

10 Thornbury Road, Plymouth PL6 7PP, United Kingdom

British Library Cataloguing in Publication Information Available

Library of Congress Cataloging-in-Publication Data

Daempfle, Peter.
  Good science, bad science, pseudoscience, and just plain bunk : how to tell the difference / Peter A. Daempfle.
    pages cm
  Includes index.
  ISBN 978-1-4422-1726-3 (cloth : alk. paper) — ISBN 978-1-4422-1728-7 (ebook)
(print) 1. Science—Methodology. 2. Science news—Evaluation. 3. Communication in science. 4. Science—Study and teaching. I. Title.
  Q175.D174 2013
  500—dc23

                                                                    2012034880

∞™ The paper used in this publication meets the minimum requirements of American National Standard for Information Sciences—Permanence of Paper for Printed Library Materials, ANSI/NISO Z39.48-1992.

Printed in the United States of America

For my wife, Amy
For my children, Justina and Konrad
For my father, Tobias

Contents

List of Figures and Tables

CHAPTER 1

CHAPTER 2

CHAPTER 3

CHAPTER 4

CHAPTER 5

CHAPTER 6

CHAPTER 7

CHAPTER 8

CHAPTER 10

CHAPTER 11

Preface

This book shows the reader how to think like a scientist. It defines what science is and how it can be misused. Throughout the book, provocative scientific examples are provided that guide the reader to consider publicized claims more critically. The tools to question authority and think scientifically are given by exposing the reader to scientific cases and contemporary studies. Each example provided walks the reader through the research methodology and mathematics, history, philosophy, and educational research needed to understand the science. In this way, the roots to science literacy are acquired. The excitement and optimism science promises for the future is juxtaposed with threats to scientific integrity and the declining results of national science efforts.

Science is shown, throughout the book, as a dynamic part of a changing society. In short, this case-based, interdisciplinary approach to learning about the scientific process leads the reader to real truths behind the many natural phenomena treated. *Good Science, Bad Science, Pseudoscience, and Just Plain Bunk: How to Tell the Difference* develops an appreciation for the way in which we gain scientific knowledge. Debunking science myths and pseudoscience helps the reader to learn science information in a fun, applicable way.

Science standards are addressed by applying scientific information within cases, controversies, and history. The purpose of the book is thus to excite the reader's innate interests in the scientific process and make the public better consumers of scientific information.

Outstanding or unique features of the work include:

- Original material from my research on the retention of students in undergraduate science, scientific reasoning, and the transition between high school and college science. The research has been peer-reviewed and published in separate articles in various refereed journals.

- Unique real-life examples, stories, poems, and ethical dilemmas that are provocative, held together by a concise overview of the aspects of scientific literacy.
- Quotations from respected scientists that bring the reader into the mind-set of science.
- Debate and discussion stimulations for instructors using the text presented by tapping interesting aspects of science, philosophy, history, mathematics, education, research methods, and critical-thinking strategies.
- Help for consumers of science to anchor their own thoughts and/or to provide teaching strategies for one of the many topics in each chapter of the book.
- Models of reasoning and guidelines for thinking critically.
- Information to make science accessible to a broad range of audiences and yet touch all of the areas needed for a full understanding of science.
- Major themes common to all scientific disciplines presented in a clear and readable manner for the general public.

Good Science, Bad Science, Pseudoscience, and Just Plain Bunk: How to Tell the Difference emphasizes an approach to learning science that is based on science as a process. Beginning chapters discuss process in a brief manner, with a focus on the analysis of an experiment. The book is an extension of the historical, philosophical, sociological, and mathematical ideas of science, which gives a full flavor of the underpinnings that compose scientific thought.

Cultural components to STEM (science, technology, engineering, and mathematics) success are discussed in order to determine a plan for improving access to these fields. Research on STEM education, careers, and social issues contributes to a unique strategy for success in U.S. retention and recruitment into STEM fields. The book advocates social pressure to move STEM to a forward position in our national priorities.

Current trends show a need for an understanding of science process increasing in most areas of the economy. Fewer professional jobs exist without the need for the knowledge of skills in science areas. The major textbooks only dedicate a piece of their books to such learning because of the need to cover content. This work infuses recent economic developments with changes in STEM areas. The ability of STEM fields and STEM education to drive future economic growth is explored. While there is a focus in the first chapters on provoking debate to stimulate the reader, this book continues to help readers detect good and bad science throughout the chapters.

The main audience for this book is people interested in science or science education. This includes readers who are entering STEM teaching or have children entering college science courses, people with a nonscience background who want to gain greater understanding about science myths and topics, and those who would like to gain insight into the state of science education today.

This book is also useful in establishing the foundation of course content for traditional, online-only, and hybrid introductory science methods courses. Other courses in which this book may be useful include: science research methods, introduction to education, science education reform, advanced high school science/honors courses, interdisciplinary science, integrated science, and critical thinking in science.

Those who want to increase a reorientation toward a STEM focus in the core curriculum of postsecondary institutions will find this book instrumental in such efforts. Instructors who are developing STEM courses for nonmajors to draw them into the related fields could use this book as a central text.

The integration of interdisciplinarity across the curriculum and within the sciences is reflected in the content of this book. History, philosophy, mathematics, science education, sociology, and politics anchor the book within a scientific framework, which is necessary for collateral thinking.

The use of this book in the often idiosyncratic interdisciplinary science courses allows the liberal use of everyday topics to augment the book and to personalize the course for each instructor. This book is broad enough to allow for flexibility in course design but sets forth a valuable core curriculum. Extensive references allow the instructor to research and integrate course materials with their own research articles. For instance, chapter 3, "Tools Scientists Use," discusses the importance of statistics in evaluating research, and an instructor could augment the chapter by bringing in external articles for students to critique. The background from the book will provide the tools to conduct the evaluation.

MY BACKGROUND

My background and interests place me in a unique position to produce a book designed for people who are interested in learning about how to interpret and use science or who want to use science in their particular teaching field. As an educator for twenty years in postsecondary and secondary science classrooms, I know many of the questions and misconceptions people often have about science. I see the challenges to success in STEM courses and majors.

I hold a PhD in science education, which is tailored to studying what constitutes science literacy and how to reach a level of knowing in science that makes the public better consumers of science products, both in terms of research and tangible technological objects. I worked as a science advisor and consultant to develop and implement science content and policy for the No Child Left Behind Act for eight years to improve science standards and assessments and have researched and published on the transition of high school students to college science programs. Even graduates of science programs are often unfamiliar with the interdisciplinary excitement that science offers and the methods of the scientific process. People need a book like this to give them a reason to think like a scientist. *Good Science, Bad Science, Pseudoscience, and Just Plain Bunk: How to Tell the Difference* will bring readers into the scientific paradigm.

Acknowledgments

This book emerges from my textbook *Science and Society*, which is geared for college students entering introductory scientific study. *Good Science, Bad Science, Pseudoscience, and Just Plain Bunk* expands upon its ideas of skepticism to analyze societal and cultural impacts on STEM advancement.

I thank my wife and editor, Amy E. Daempfle, PhD, who is my inspiration. I thank my father, Tobias Daempfle, who has had great impact on my life and whose conversations have contributed significantly to this book.

To those who have died and to the generations that have come before: Justine Daempfle, my mother, and Josefa Nick, my grandmother. May there be a better place to meet again.

I thank all of the scientists and science educators, whose contributions are often unrecognized but part of a greater whole. I hope that this book will invigorate a passion and appreciation for their works.

I also thank my teachers of the past, who helped form the foundations of thought to create this book. Their ideas were interdisciplinary, creative, and forward thinking. Their teaching was profound. They gave a great deal, more than they can know: Carole Demian, Robert F. Pospisil, John P. Rosson, Wendell W. Frye, William M. Elliott, Audrey B. Champagne, Robert F. McMorris, and Margaret Kirwin. Let their teaching touch eternity. To my students, whom I have had the honor of teaching and sharing ideas: let our paths cross again through this work. And to those editors at Rowman & Littlefield who made this publication happen: Sarah Stanton, Jin Yu, and Elaine McGarraugh.

Finally, I would like to thank my colleagues, students, and administration at SUNY Delhi who have been in support of my writing and research through intellectual discussions, professional development grants, and a sabbatical. In appreciation.

I

SCIENCE TOOLS

1

Introduction

Science bombards our everyday lives. Whether through photons of light from an incandescent bulb developed over a century ago or a life-saving medical technique helping hearts, the way in which science changes our world is obvious. Our culture is now so dependent on science and its unique characteristics. Yet few people have the tools to understand how science works and how it has progressed so far in such a short time.

Thus the main purpose of this book is to show the reader how to think like a scientist. It defines what science is and how it can be misused. Throughout the book, provocative examples are provided that guide the reader to consider the facts behind a scientific statement more critically. To illustrate, consider the claim "Cigarettes do NOT cause lung cancer." Most readers are surprised at this assertion, but there is data that supports it. In a study of over one hundred thousand smokers versus nonsmokers, there is no statistical difference in lung cancer rates between the two groups, given rigorous statistical tests. In addition, what percentage of smokers actually gets lung cancer? One may guess 10 to 20 percent. Actually only 1 percent or 1 in 100 smokers actually develop lung cancer. These are, in fact, valid statistics used by tobacco companies for decades. So, light up? Before you do, contemplate the opposing set of data.

What percentage of people with lung cancer were once smokers? It is quite high (90 percent) and statistically significant. Nine out of ten people on the lung cancer floor in any hospital were smokers. When the numbers are presented in this manner, the link between smoking and lung cancer is clear. Smoking is very much linked to lung cancer. Many people were tricked by the first set of data presented by the Tobacco Institute and Council for Tobacco Research in the 1950s, as shown in figure 1.1. That resulted in addiction to nicotine and in unnecessary cancer deaths over the past century. Obviously, research and mathematics can be deceptive and used to

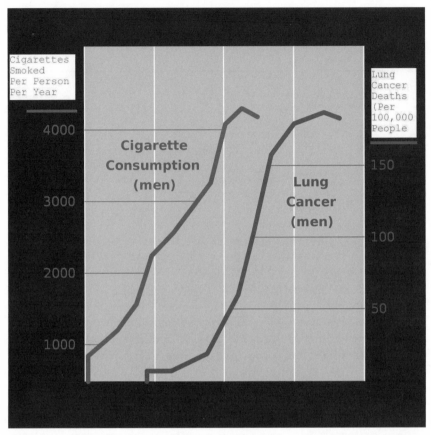

Figure 1.1.   Graph of year (x-axis) versus smoking and lung cancer incidence. Source: www.nih.gov.

manipulate decision making. Both sets of data are true, but each presents a different picture. The first section of this book gives readers the tools to interpret science research information and to form their own conclusions.

Science is based on good argumentation to discover the truth about a natural phenomenon. The philosophical foundations of scientific reasoning are given early on in the book, with original research as a foundation to use the tools to engage in science. The book shows the importance of skepticism and the uncertainty inherent in scientific thought. Is it better to simply accept the facts in a newspaper or website or to critically evaluate evidence? Should one just accept what the doctor says in a medical situation? The book provides the skills to question authority just like a scientist. Everyone, not only science people, need to practice this—it may be life saving.

To do this, foundational science tools such as mathematics are treated through the application of provocative cases and arguments. The book entreats the reader to enter into a world of the past, joining the struggles of scientists from before the scientific revolution through the present. Consider the plight of Socrates, who was sentenced to death by hemlock for questioning the ancient democratic Athenian government. In essence, the book inspires its own scientific revolution within the reader, awakening a reaction against a groupthink mentality.

The excitement and optimism science promises depends on this way of thought: thinking independently. The book helps readers to do this by showing them the many facets of analyzing science. Humankind's rapid advances within a short period of geologic time are evidence of science's optimism. What are the next great advances? We do not know because they have not yet been developed, but the reader is the key to the future. In a world taken in by the loud and the superficial, reason and science are the only ways to truth. Questioning is the first step to "becoming" a scientist and an independent thinker.

Focus is aimed at a variety of threats to science and thus the civilized world. It comes at a delicate time in our nation's history. Challenges to scientific integrity, serious roadblocks to science education, and the lure of competing pseudosciences threatens our nation's science literacy, competitiveness, and security. Recent polls show high support for questionable and untestable phenomena, with 50 percent of people believing in ESP (extrasensory perception), 42 percent in haunted houses, and 34 percent in ghosts—but less than 5 percent can explain why our cells need oxygen![1]

These numbers are not surprising given the lack of science role models in society—intelligent people are depicted as "nerds," and sports heroes, Hollywood, and fads are given artificial elevation. This is not to impune people who like movies and sports, but it underscores the shift in respect from intelligence and productivity to appearance. These trends contribute to shortages of science-oriented career professionals and threaten medical care, science research, teaching, and military development. According to the U.S. Department of Labor, only 5 percent of U.S. workers are employed in STEM (science, technology, engineering, and mathematics) fields, but these fields account for 50 percent of our economic growth.

In fact, the country's fifteen-year-olds rank seventeenth in the world in science and twenty-fifth in mathematics, our infrastructure ranks twenty-third and our life expectancy twenty-fifth, but we are number one in obesity rates and debt according to the Organisation for Economic Co-operation and Development. A lack of focus on scientific thinking and reasoning has led to some poor decisions.

An urgent change in culture is recommended before it is too late. Original research on science teaching and retaining students, with suggestions for improvements, will be given to reverse these deleterious trends. It starts with a better knowledge of science. The reader will be shown how to organize the massive knowledge base and to explore the underlying themes that emerge from the content through analyzing cases, pseudoscience, and science myths. For example, full appreciation for what is happening during an earthquake draws from the physics that underlies the motion of plates, the history of where energy is coming from in the universe, the biological effects on people at both medical and evolutionary levels, and a vision of the chemical nature of Earth's structure.

Recognition is also given to our limits of understanding and our battle to control nature. The greatest struggle of all, of course, is not the natural world around us but the nature within—our struggle to control ourselves. It is human nature to want control over our own lives and to have "free will." Are we merely a tiresome set a genes, controlled by our inherited chemicals? The nature of the genetic basis of behavior will also be explored.

Science is shown, throughout the book, as a dynamic part of a changing society. In short, an interdisciplinary approach to natural studies leads readers to discover the truth behind their unanswered questions. Perhaps this book will excite new interests in science. Its main goal is to recruit good people into becoming their own scientists. Our society needs them.

SCIENCE LITERACY

Understanding science requires our citizenry to be scientifically literate. Science literacy is defined by several government documents as a body of knowledge, an appreciation and a set of skills necessary to explore science and the scientific process. Understanding how and why things work the way they do in the natural world is an important part of science literacy. But how many of us actually know the often very intricate and exciting components of a scientific system? According to recent international assessments of student learning in science by NAEP (National Assessment of Educational Progress), TIMSS (Trends in International Mathematics and Science Study), and PISA (Program for International Student Assessment), the results are dismal. Most U.S. citizens are operators of products of science but do not understand the equipment or the process of producing science. After all, almost everyone uses a computer, but how many are able to explain how it actually works?

Further, who should acquire science literacy? Should our government's goals be that every citizen achieves a certain level of science knowledge? Is this objective even possible? Currently, the trend in standards-based school assessment is to have every person reach a level of science achievement under the No Child Left Behind Act of 2001 and Race to the Top of 2009. Both administrative plans purport that all students can universally learn science to a level that will help them make life decisions better and function in modern employment. Learning science and the scientific process is a noble and humane goal for improving peoples' lives, a focus of this book.

National standards and assessments have not always worked toward educating everyone. For example, the 1969 NAEP goals were to educate those that were capable of being taught science. In older standards-based educational documents, a theme emerges contending that not everyone can or should be educated in science.[2] This is not acceptable by today's governmental, educational, or societal outlooks on the role of science education. In chapter 2, "Science Is Arguing," we will explore the levels of reasoning and argumentation that are required to successfully engage in science. We will also review the educational philosophies on the nature of knowing to determine whether or not our modern goal of educating everyone is possible or even desirable for a successful twenty-first-century society.

SCIENCE IS INTERESTING

Throughout this book the reader will tour the many facets that make science both fascinating and yet different from any other way of thinking. The latest scientific advances are tapped to help readers explore how science is done. There is a great deal of reliable and sound scientific information publicized, but with the explosion of knowledge transmission via technology, many untruths are also easily propagated. Myths, pseudoscience, and outright lies pervade science explanations given to the public. By looking at different aspects of scientific thinking, the reader gains the tools necessary to evaluate and properly understand the science presented every day by the media.

Historical, philosophical, mathematical, and social underpinnings of science are often ignored but are probably the most important part of any scientific information presented. These aspects will be traversed to help show the reader just how science knowledge is created. This book is not a "how-to" book with step-by-step directions on becoming a scientist. Instead, a series of stories, poems, research studies, vignettes, and cases are presented and infused with the many roots to science that make it come alive.

Unfortunately, the media and even schools present science without a base for how that information is discovered. To illustrate, people are often told that "exercising is good" for them to gain more muscle and improve general health. Few are exposed to deeper science happenings of muscle development. Most do not know that new muscle is never actually created; your body produces a certain number of muscle cells and that number cannot be increased through exercise. However, when one exercises, muscle cells do grow larger and their smaller subunits, called mitochondria, divide. Those are special, cigar-shaped powerhouses of the cell, generating energy by using food, such as bread, meat, vegetables, and fruits, in addition to Ho Hos and bacon, to burn off calories. A mitochondrion is shown in figure 1.2.

In fact, mitochondria divide on their own because they are self-contained units that are semi-independent from our cells. Mitochondria have their own DNA (deoxyribonucleic acid), and their own structure, and they are thought to be ancestral bacteria. These organisms were absorbed into larger cells and propagated, both obtaining some benefits. The mitochondria acquired free housing from the larger cells but also had the chemical ability to give energy to their new "homes." This made the new combination a more effective and competitive structure. This is how the start of modern cells, called eukaryotes, began.

Mitochondria are inherited from mother to child because the egg of a female contains cell structures or organelles (such as mitochondria) and sperm contains only genetic information. A mother's mitochondria are thus handed down from

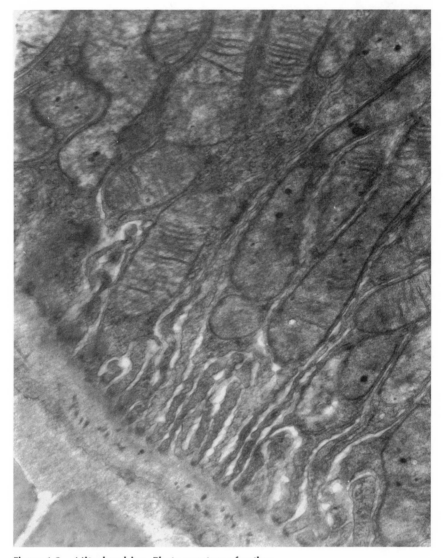

Figure 1.2. Mitochondrion. Photo courtesy of author.

generation to generation so that a little piece of "Mama" is in every one of us! This unique set of genetic material is tested during crime scene investigations to aid in the identification of potential criminals.

This view of mitochondria is called the endosymbiotic theory and was developed by Lynn Margulis, who studied mitochondria to help explain how our modern cells evolved. It is difficult for media sources to educate people on all of these issues, but the underlying information is what makes the topic most interesting.

FOUNDATIONAL SCIENCE

Science is complex, so let us begin with science basics: the methods and knowledge to successfully facilitate a scientific way of thinking. These foundations are so vitally necessary for interpreting research reports that emerge daily in our media. For example, a UCLA (University of California at Los Angeles) study was reported in October 2011 about the role bacteria play in causing pancreatic cancer, a dreaded disease in which only 5 percent of people survive after five years. The report shows that *Streptococcus mitis*, a type of oral bacteria, declines in numbers in the oral flora of people with pancreatic cancer. *Streptococcus* is a genus of bacteria that contributes to tooth decay and looks like chains of circles under a microscope. The report states that tooth decay is thus linked to pancreatic cancer and concludes that "good hygiene can avoid the terrors of getting the disease."

Obviously, the report is erroneous in its conclusion because it does not match the data. People with pancreatic cancer have fewer *S. mitis* cells in their mouths, so effective tooth brushing would have the opposite effect—increasing pancreatic cancer susceptibility. So why does the report make such a claim? It is probably because such a finding sounds good. By making people feel empowered because they could somehow avoid pancreatic cancer by brushing, the report is positive, interesting, "headline grabbing," and makes sense to the public. However, it is an example of bad science in every way.

Not only is it wrong, but no information is given about the methodology, experimental design, or mathematical backing behind the data. In fact, upon further investigation beyond the news media to other more rigorous science journals, it appears that only ten subjects were tested in the study and no real claims could be made. The sample size of ten is very small and lacks the statistical power to support generalizations made by the media, which called for national testing of people's mouths for the disease.

Research reporting such as this bounces the public back and forth with weak research. A friend of mine gave up coffee for three months because of a report citing coffee and pancreatic cancer links. The deaths of Steve Jobs of Apple and Hollywood star Patrick Swayze heightened people's awareness of pancreatic cancer, stimulating a rash of reporting in 2012. With the knowledge that such claims will get attention, the media is all too willing to make headlines from the data. The media's inability to interpret the numbers behind the science demonstrates the need for improving the mathematics literacy underlying science literacy.

In chapter 3, "Tools Scientists Use," science is discussed in terms of the need for rigorous methodology and the necessity of statistical mathematics to give power to the research. The ways by which science knowing happens—the scientific method, uses of and processes of experimentation, the role of science modeling and nonexperimental science—will be explored. These tools help empower the reader to interpret real and false claims made by both scientists and the media.

DEFINING SCIENCE

Science is an interrelated paradigm with concepts and terms branching from non-science domains as well as related science fields, as described in chapter 4, "Science for Every Person." This chapter introduces readers to the culture of science education and learning. With the explosion of information in science, it is no longer possible to merely deliver it all to people. Instead, the chapter shows the reader how to organize that massive knowledge and see underlying themes that emerge from the content. It is an important chapter for easing parents and rising college students into the transition to college science. An overview of science themes is provided in this chapter to give a foundation for the book so the reader can form greater conceptual understanding and place the information into meaningful networks. The cultural transition to studying science is discussed along with methods of learning science. Science knowledge is seen as a whole, with separate parts contributing and interconnecting, as depicted in figure 1.3.

Science is often seen as separate entities, with biology, geology, physics, and chemistry separate but related, with overlaps into one another's areas as well as with mathematics and society. Science occurs within a richness of knowledge linked to all other areas of study. The distinction between areas of thought found in every educational institution was well discussed by the geographer Halford Mackinder in 1887, who stated, "The truth of the matter is that the bounds of all the sciences must naturally be compromises, knowledge . . . is one. Its division into subjects is a concession to human weakness."[3]

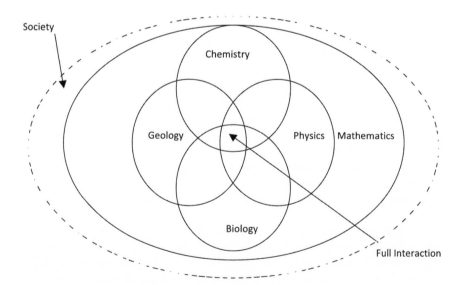

Figure 1.3. Science knowledge seen as a whole.

Unfortunately, a definition of *science* is difficult to develop. Even scientists have trouble agreeing on a meaning because there are so many facets to the word *science*. It is a way of knowing—a way of thinking about the world. It is a process by which we understand the universe. It is, in its purest form, a way of arriving at some truth. But *knowing, understanding,* and *truth* are all hard to define.

A modern definition of science consists of three parts. First, it is a *body of knowledge* about the natural world. There are known facts and concepts on which knowing science is based. Second, science is a *method*, or a way of finding out the truth about something in the universe. The scientific method has required elements that make it different from other nonscience explorations. Third, science requires *reasoning* or *critical thinking* to understand the base knowledge and to apply new discoveries derived from using the scientific method. Scientific reasoning takes skill and practice. It is infused throughout the book, but its importance and strategies for development are treated in chapter 5, "The Role of Critical Thinking." It comprises rationalism (a link between cause and effect), openness to investigation and indefiniteness, and an ability to recognize patterns and solve problems. Reasoning is important in everyday life and is the foundation for scientific advancement.

Science literacy is even more difficult to define, but it is generally the facts, skills, ideas, and ways of thinking that enable a citizen to make decisions about and use science and technology. Everyone should be scientifically literate. One does not need to be able to perform specified tasks or understand complex instrumentation to be scientifically literate. Science literacy is very different from doing science. Scientists are specifically trained to be experts in certain domains. The goal of our society is not to make everyone a scientist, but to get everyone to think like a scientist: that is the best descriptor of a scientifically literate person.

HOW TO THINK LIKE A SCIENTIST IN OUR SOCIETY

The reader is invited to engage in a treatise on how scientists think. This book illustrates the role of science in leading human progress but cautions on the dangers of not having a scientific outlook from an individual level to national and global threats. For example, chapter 6, "The Media," depicts how science facts and conclusions are made accessible and inaccessible to the public. Examples expose how easy it is to make erroneous decisions based on media bias, inaccuracy, and poor methodology and foundational science. The reader will become savvy to the pitfalls of misinterpreting and misusing science data based on his or her use of the readings from part I, "Science Tools."

To illustrate, consider the following report on the major U.S. news networks in October 2011. The "discovery" was made that "people who have more friends on Facebook have better developed regions of the brain."[4] The news reports claim that Facebook socializing develops portions of the brain that result in better skills in social functioning. A simple relationship was made—without data on who funded the research, how many subjects were involved in the study, and what types of measures were used in the so-called developed areas of the brains under observation. The media can be very useful in getting science information out to the public in an accessible manner, but the obvious omissions mentioned above lead one to question whether it is a good means of scientific transmission.

What drives the media to make such claims to the public? A commonly cited explanation is that there is limited space or time available for news organizations to give all of the facts; if a person is interested in knowing more, he or she can research the topic. Given the poor science literacy skills reported by international assessment measures, this assumption of public ability is irresponsible. There must be more to the media's poor reporting of science. After all, sports reports and articles give far more data and have more time devoted to them than science research findings.

Reporting the kinds of relationship interpretations in the above example of Facebook-related brain development may show both political and economic agendas to push technology—a kind of groupthink—to save the U.S. economy. If the public accepts that Facebook is good for it, then the part of our gross domestic product (GDP) dedicated to Internet commerce and media exposure online will strengthen. Or is it simple incompetence in reporting? Either way, we will explore the role of the media as a help and hindrance to scientific thinking and attempt to recommend changes based on our findings.

Of course, the media's and educational systems' failure to properly promote science literacy has manifested many rumors and misconceptions in almost every area of science. In chapter 7, "Pseudoscience," and chapter 8, "Debunking Science Myths: Separating Fact from Fluff," the reader learns about science while debunking the information presented. Numerous science myths and false sets of information prevail in the absence of science literacy. These chapters allow the opportunity to apply the tools of scientists to analyze the many cases and studies that form our science conclusions. Each topic gives the science behind the myths and mistakes in accepting information. Examples range from the role of high-fructose corn syrup in obesity and diabetes, to the evolution of sexuality and human kindness, to whether humans cause earthquakes. It is a set of chapters that should entice us to continue our own research into our own personal science interests.

SCIENCE AS A UNIQUE COMMUNITY

The science community needs to stick together in the face of the many threats to its pure search for truth. Chapter 9, "What Are Scientists' Responsibilities?," acknowledges that science research, as seen in the examples of the previous chapters, does not exist in isolation. Scientists communicate and contribute in a subculture to produce advancements, but science is actually a social endeavor, and scientists form their own societies. This chapter discusses scientific thought and innovation with consideration for historical and social contexts and their influences. Thus, the inherent question of what responsibility the scientist has in a society will be debated. The media and the scientific community are dependent upon each other, and yet there is an innate mistrust. What should the relationship be? How can the public best benefit from changes to both?

The scientist has a role in shaping the world and a responsibility to oversee information that could end life as we know it. The chapter can be used to address the possible arguments about technologies produced by scientific advances and about some of the dangers faced by the public. Recombinant DNA techniques, for example, use bacteria and other organisms by inserting genes to make products for humans. These products have both benefits and potential dangers for the public. A gene for a particular drug is inserted into the DNA of a bacteria or plant, and as the organism grows, the product is made. It is efficient and has led to the creation of many transgenic organisms, or those with genes from species other than their own.

A major change in medical treatments, human growth hormone (HGH), has been produced this way for the past twenty-five years to help extreme cases of growth disorders. Before recombinant DNA techniques, HGH had to be extracted from the pituitary gland of cadavers and had risks of contamination to patients from the dead material. Now with the new techniques, HGH is fast and easy to produce, without contamination risks, making it more commonly available. Should a preteen male, predicted to grow to a height of five feet four inches, take the drug against the doctor's advice? What are the side effects? What are the social issues involved in being short? Will another doctor help the patient if one doctor denies treatment? What is ethical for the doctors to do? These are just a few questions about HGH that are provocative and yet face thousands of young Americans every day. Many cases such as HGH will be addressed in this book to help the reader weigh the pros and cons of using technologies that scientists have made available to them. The role of the scientist in developing products for society will be discussed in more detail in the book. Historical and contemporary figures' opinions, along with research and case studies, will help clarify the ethics of scientific progress.

Obviously, a scientific way of thinking is threatened by forces in society that are as old as history. The last part of the book, "Science: Threats or Compromises," treats some major shifts in societal thinking that point to a recent history of rapid advance in science products but also to a decline in science literacy. In the absence of critical thinking and the individuality inherent in science literacy, humans have turned to groupthinking and even violence to answer their questions regarding "What is truth?" Albert Einstein, on concern for a rise in restricted freedom of thought in the totalitarian world of the era, made a point of this in the following poem:

> By sweat and toil unparalleled
> At last a grain of the truth to see?
> Oh fool! To work yourself to death.
> Our party make truth by decree.
> Does some brave spirit dare to doubt?
> A bashed-in skull's his quick reward.
> Thus teach we him, as ne'er before,
> To live with us in sweet accord.[5]

The ramifications of this groupthink mentality mark a significant threat to modern scientific progress. Individuality and skepticism drive science and creativity to unlimited bounds, as seen by Einstein's discoveries. The push for individuality is a main theme of this book. Science helps to keep our minds open and to think, as Einstein did, about phenomena without restriction. However, people taken in by the popular and the superficial will not be allowed to grow intellectually and will be flummoxed by the myriad of false sciences.

AN AGE OF OPTIMISM

In chapter 10, "Science Progress and Challenges to Science," the miracle of science and how rapidly it has progressed within such a short time period is discussed. Consider the age of the universe, estimated to be about fifteen billion years. In terms of geologic time, humans inhabited Earth only at the very end of this timescale, no more than ninety thousand years. A single individual's life span is no more than a century. Yet a person can accomplish so very much in that short time frame. Aristotle, Sir Isaac Newton, Galileo, Marie Curie, and Einstein produced a wealth of scientific change and knowledge during their lifetimes. The accomplishments of scientists are important to understanding scientific progress. However, the threats to future contributions in science will be juxtaposed in the latter portion of the chapter. Of all the varied threats to future science progress and individual science thinking, an inability to critically consider information remains the most virulent threat to our progress. Humanity requires at least a subset of individuals to cope with the many challenges humans must face, from earthquakes to diseases. Without a group to think like scientists, our fate is hopeless.

Consider that we, as humans, are physically weak; the dinosaurs would have been able to devour us years ago (if we had existed along side them). Our minds are our strength. In 1665, at the end of the scientific revolution in Europe (1540–1690), mathematician and scientist Blaise Pascal (1623–1662) retired and began writing his thoughts: "Man is but a reed, the most feeble thing in nature; but he is a thinking reed . . . All our dignity consists, then, in thought . . . by thought, I comprehend the world."[6] This quote reflects the frailty of the human condition but also the strength of human thought. Pascal's words embody the essence of the emergence of a way of thinking about the natural world. It was a movement from the power of physical strength to the power of the mind. The scientific revolution comprised a period of skepticism about existing knowledge about how the world worked. This book seeks to reemerge a national thinking that will recapture the excitement of the scientific revolution.

IS SCIENCE EDUCATION THE ANSWER?

Is the educational system able to seize this opportunity upon our need to advance science to all people? The results of the international studies mentioned earlier in the chapter indicate that No Child Left Behind has not improved the United States's international standing among nations, with our science and mathematics results comparable to those of the developing world. The United States spends more money on education than any other nation on earth per capita. Why such miserable results in science and mathematics achievement? We will explore the nature of science teaching further in chapter 11, "Getting People to Love Science," for some possible answers and recommendations for improvement. The disjuncture between science teaching and science learning is a national problem that needs to be successfully addressed to further scientific literacy and national progress. We will also explore the aspects of our culture that work for and against STEM successes.

One problem is that too few college students are recruited and retained in science programs to meet the nation's future needs. The National Research Council indicates that first-year college science student success rates remain stubbornly low. More specifically, first-year student retention in STEM majors were at levels less than 50 percent from 1977 through the present.

THE HIGH SCHOOL/COLLEGE SCIENCE DIVIDE

U.S. census data also show that potential majors in STEM areas are lost particularly in the transition from high school to college by first-year students switching from STEM majors to non-STEM majors.[7] Researchers in science retention indicate that the highest losses from STEM majors, on the whole, occur at this juncture (between high school and the first year of college). Losses range from over 50 percent in the biological sciences to 20 percent in the physical sciences and mathematics.[8]

The greatest losses of students were found among high school graduates who withdrew their decisions to enter a STEM major at or even before enrolling in college.[9] The college-level instructor can do little about the loss at this juncture. However, during college, the highest risk of switching from STEM fields (35 percent) occurred at the end of the first year.[10] As a student's time in college increases, however, retention of that student improves, with reports of a loss between sophomore and junior year of only 2 percent, and from the start of the junior year to graduation, 0.8 percent. Students become vested in scientific thinking, and we need to explore that development further. Interestingly, however, very few students transfer into STEM majors after college enrollment, and there is always a net loss.[11] One researcher pointed out that "not only do the sciences have the highest defection rates of any undergraduate major, they also have the lowest rates of recruitment from any other major."[12]

Career entry in STEM areas has been adversely affected by these attrition rates, whereby both the health professions and engineering areas lost over half of their entrants (53 percent and 51 percent, respectively).[13] Considering the rise in college enrollment, the reported decline of 60 percent in the number of students preparing to teach science is alarming. Although a variety of factors have purportedly contributed to this situation of STEM teacher scarcity, the poor retention rates for STEM majors have clearly played a role.[14]

The decrease in STEM enrollment is coupled with a declining scientific literacy of the population as a whole. This transformation has produced a nation that has "simultaneously and paradoxically both the best scientists and the most scientifically illiterate young people."[15]

STEM DECLINES AS AN ECONOMIC THREAT

Declining scientific literacy, combined with the decreases in STEM enrollment, resulted in reduced numbers of qualified individuals available for research development, a driving force in the progress of science. Numbers of qualified health care professionals, especially primary care physicians and nurses, science teachers, and military personnel, have been adversely affected. These areas are generally well-paying, available jobs, and yet the declining interest in STEM areas occurs in the face of widespread underemployment and unemployment. Even a lack of opportunity for other avenues of career choices has not driven people into STEM areas.

Thus, public concern has been vocally expressed due to governmental and academic agencies reporting declines in our international competitiveness in the science-and-technology-dependent sectors of the economy.[16] The final chapter, chapter 12, "Driving the Economy through Science," shows a planned path for the United States to economic success based on educational and societal changes in order to vault our national competitiveness and improve job growth and our economy. Many economic variables depend on global STEM growth and development. Improving each person's ability to think like a scientist is in everyone's best interest.

2

Science Is Arguing

Sir Isaac Newton was a great English physicist and mathematician who gave us a central understanding of modern movement, particularly gravity, optics, and ideas about "fluxions" (calculus). Newton's reflecting telescope gave him initial fame. In 1672 he was made a fellow of the Royal Society, discovering that white light is composed of the same colors as a rainbow. He published *The Opticks* explaining these views but also published works in history, theology, and alchemy (early chemistry). His greatest work, the *Philosophiae Naturalis Principia Mathematica* (*Mathematical Principles of Natural Philosophy*), brought together ideas of gravity and universal movement of objects based on forces.

Isaac Newton held numerous honors. He was appointed ward of the Royal Mint in 1696; elected as a member of Parliament for Cambridge University (1689–1690 and 1701–1702); and was knighted and elected president of the Royal Society, an office that he held until his death in 1727. Despite his vast accomplishments, he was depressed. Newton found himself in continual argument with colleagues and friends, deep in thought about alternatives to his and others' viewpoints. While unhappy in his personal life, his argumentative strategies are emblematic of the kinds of thinking required to propel science. Through considering others' viewpoints, debating, and questioning, he may not have been well liked, but his personality facilitated great accomplishments.

Skepticism and debate are inherent in successful scientific thinking. Without questioning and troubleshooting, a conclusion might be accepted that is false or a true conclusion might be rejected in error. Skeptics are often not popular, but they are vitally important to establishing the logic of an argument and reaching the most valid conclusions possible. For example, when fluoride was first added to toothpaste in 1914, it was not supported by the American Dental Association (ADA). The idea of it was rejected by many consumers as well. I recall a granduncle, a tidy Capuchin priest well into his nineties in the 1970s, with no dental caries (cavities). He vehemently opposed fluoride toothpaste and instead brushed with baking soda. My granduncle was a skeptic, and his data point (of being caries-free at ninety-five) did not support profluoride research. The thought of fluoride, a toxic chemical, added to the oral environment was unacceptable to him.

Much of the science community was against fluoridation; thus, there was an ingrained argument. Arguments and opposing sides are needed in discovering scientific truth. In this example, fluoride efficacy was debated to determine its effectiveness by testing on human and animal subjects.

One data point alone does not constitute a scientific study. Proctor & Gamble worked feverishly in the 1950s to obtain data from many samples to show that the application of fluoride in toothpastes was effective in reducing dental caries. They conducted studies on large numbers of individuals to see the effects of flouride pastes on general health and oral health. Finally, in 1960, the ADA issued a statement approving fluoride toothpaste, reporting that Crest had been shown to be an effective anticavity (decay preventative) substance. It concluded that fluoride can be of significant value when used in a consistent oral hygiene program. The scientific reasoning required to change the paradigm of oral hygiene was composed of a combination of scientific argumentation coupled with experimentation and mathematics to support the claim that fluoride was beneficial and not harmful in small doses.

Science is constantly changing, with incentives to always improve upon past achievements. Fluoride as a toothpaste additive helps to inhibit bacterial growth by literally "sucking up" electrons from bacteria's biochemical pathways. This prevents bacteria's chemical reactions from occurring and suffocates the bacteria that cause tooth decay. Fluoride is, in fact, poison to all living systems, including humans. However, in our mouths, fluoride toothpaste exposes humans to such small doses that it is effectively harmless. Interestingly, its mode of action is quite different from recent efforts to improve outcomes for dental caries prevention.

In 2006, the biotech company BioRepair began testing the first toothpaste with the additive hydroxyapatite as a way to prevent dental caries. While accomplishing the same ultimate task as fluoride, hydroxyapatite works in a different way. Hydroxyapatite adds an extra layer to the enamel of a tooth to protect it from bacterial acid. It is found normally in all of our bones, and it remineralizes to strengthen bone material that is newly formed or added. A breakthrough to supersede fluoride's effects would change the industry. It is this kind of competitiveness and capitalistic philosophy that drives research science. Without the will to beat out the other company or to excel beyond existing knowledge, science would be stagnant. Scientific argumentation is thus rooted in competition. However, scientists also need to be able to work together well to share ideas and build upon knowledge.

Unfortunately, it seems science progresses most during times when there is intense competition between nation-states—usually in war. Wars have led to the development of many commodities we now use. These by-products of the wartime research development show how science teams compete, but collaborate within their own groups, to propel science discovery.

Consider that the American discovery of sonar to find German submarines led to ultrasound technology, used in so many aspects of medicine, from examining a fetus's body shape to checking for dangerous plaque build-ups in neck (carotid) arteries going to the brain. Radioactivity, plastics, computer circuits, dehydrated foodstuffs, and bulletproof (Kevlar) materials are all results of wartime scientific research. Competition continues to drive scientific research today. China persists in its development of high-speed rail systems, recently testing a train at 310 miles per hour. This would be fantastic for U.S. infrastructure, rivaling air transit that averages 450 miles per hour and besting our existing trains that average a slow 65 miles per hour (with multiple malfunctions).

SCIENTIFIC ARGUMENTATION

Science, at every step, is indeed arguing, based on a competition system that drives scientists to gain an innovative perspective. Innovation requires a rebellion against existing knowledge to create an intellectual shakeup. There is thus an argumentation system that acts as a foundation to scientific thought. This and the next few chapters examine our own ways of analyzing scientific issues through studying the components of an argument, the background belief system behind why a particular side is taken, and methods for thinking critically.

The textbook definition of argumentation, from a variety of philosophers, describes it as a set of premises (propositions) asserted in a certain way to establish the truth of a conclusion (e.g., Freeman, 1988; Govier, 1992).[1] However, in scientific thinking this characterization is too rigid for the informal logic required to develop investigations. The definition does not take the context of the argument into account. Because knowledge is created through an argument within a dialogue, reasoning about an argument needs to take place within the context of a dialogue among people. The development of BioRepair, for example, will not take place in isolation; a scientist needs other people's research and ideas to build upon in order to develop new hypotheses and statements.

Argumentation in science is not quarreling, as the term *argument* may imply. Aristotle first brought up this point, identifying the key elements of an argument: justification and refutation, as necessary to expose truths and not merely squabble to prove an opinion. To solve any problem, an argument posits a point and then an opposite counterpoint. Playing the devil's advocate is a necessary element to scientific thinking. While Newton (and many of my own relatives) may argue a point and are ostensibly disagreeable, it is through this kind of back-and-forth that issues are thoroughly considered. There is really no proof in science, though—only disproof of alternate hypotheses or ideas.

Postpositivist Karl Popper (1902–1994), who wrote many treatises on the philosophy of science, first presented this philosophy. He claimed that no hypothesis is ever proven because a new piece of data or idea could be found to discredit it. Only a process of elimination of alternate ideas furthers science. This process is known as *falsification*. Popper advocated the view that there is only progress in science via falsification of existing knowledge.[2]

Indirect proofs are often used in science, in which a *reductio ad absurdum* method is used. These assume the opposite of a conclusion and then show it to be a contradiction. For example, Galileo considered the false idea that objects fall to earth at different rates depending on their mass, with heavier objects falling to earth faster than lighter objects. In order to show the idea's self-contradiction, he imagined two objects of different mass tied together by a length of string. On one hand, the lighter object is imagined to hold back (slow) the fall rate of the heavier object. On the other hand, the combined mass of the tied system is heavier than either object individually, thus lending to the idea that the tied-together system would fall even faster than the heavy object alone. This demonstrated a contradiction in logic, thus disproving the idea that an object's mass was a determining factor in its rate of fall. Galileo's hypothesis, that all things fall to the ground at the same rate, was supported by disproving the opposite idea, that mass determined the rate of fall.

Argumentation is a means of persuasion, described by contemporary philosophers as far back in history as Aristotle (e.g., Govier, 2004; Walton, 1996).[3] More formally, it is a sequence of inferences joined together. But what bonds the ideas of a scientific argument? The elusive word: *reasoning*. Reasoning is done before, during, and after an argument to make it logical. It is the thought process of argumentation. Reasoning occurs within an argument and within the content of a dialogue. There is no distinction, for the purposes of this book and science logic, between the terms *reasoning*, *critical thinking*, *reflective judgment*, *premise*, and *inference*. These terms are hot-button buzzwords, and many people, even scholars, do not know the fine distinctions between them. It takes reasoning to effectively argue a point to draw valid conclusions.

However, to function effectively and within the scientific world of thought, the nuances are important. The scientific argument has certain key features. Walton (1996) provides a simple definition of an argument that contains three elements: first, that arguments are sequences of propositions in which some are inferred from others; second, that there is some issue to be settled; and third, that reasoning is used to settle these issues. An argument falls apart when any of these characteristics is compromised. Thus, when a person is illogical, cannot reason, or cannot consider key points in a counterargument, the debate is useless.

Argumentation begins with a scientific statement. Scientific statements are actually arguments taking some particular position. "It is better to brush your teeth in the morning rather than at night" and "It is unhealthy to brush your teeth because the enamel will wear away" are scientific statements that can be tested to determine the veracity of the claim. The statement is either supported or not by tests, results of the tests, and thinking about those results. Scientific statements are then testable positions. Scientific argumentation is a form of intellectual persuasion to convince not only the scientist but also a community of scholars to believe the truth about a matter.

Scientific statements are very different from casual statements about phenomena. They start a way of thinking that is based on testable procedures that lead to answers. Many casual statements are scientific in origin. A child may ask, "Why does snow fall to the ground in winter?" which leads to a base of knowledge to answer that question. By examining testable questions such as "How does temperature affect water?" and "Where does freezing take place in the layers of the atmosphere?" knowledge is derived.

Scientific thinking is thus based on argumentative reasoning. Reasoning is defined as the process of evaluating the evidence within an argument to accept, reject, or suspend judgment about a particular claim. The process by which an argument is constructed is termed *reasoning*. There are a variety of levels of reasoning an individual operates within to draw scientific conclusions in everyday life.

Consider a nutrition label on a box of cereal. When purchasing the box, many people will read the caloric count, the vitamin list, and the macromolecule composition as well as the ingredients. There are many claims made in the media about nutrition and diets, and the consumer considers these. Evaluation is actually a complex process involving prior knowledge of biology and chemistry of foods; research and mathematics behind the studies on diets and health; and judgments about the claims made both on the labels and in the food science discipline. Without a real understanding of the words on the food labels, reading the label is actually a useless task that most people give up on.

Most people are deterred from comparing food labels by reading long, hard-to-understand words in the ingredients portion of the label. Not understanding what is going to be put into our bodies can be unsettling. Consider bisphenol A, a chemical found in many plastic bottles and linings of food-storage cans. According to the *Journal of the American Medical Association* in 2011, Harvard University researchers showed that people who ate canned soup for five days in a row had their urine levels of the chemical bisphenol A spike 1,200 percent compared to those who ate fresh soup.

Bisphenol A disrupts the endocrine (hormone) system, particularly the reproductive processes, with fifty micrograms per kilogram of body weight in animals shown to affect the development of reproductive organ functioning. It is uncertain if the reproductive effects of bisphenol A are similar in humans. However, the chemical has been linked in numerous studies to human cardiovascular disease, diabetes, and obesity.

Spikes in bisphenol A are eliminated through the urine and are considered temporary. It is thus argued that the impact on human health is minimal. The U.S. government's health and environmental agencies continue to allow foods to be packaged with the chemical despite the possible harms. Further action and research is needed to better establish the link between human health and bisphenol A. Finding alternatives to packaging foods requires money and time and may inhibit economic growth of the industry.

In contrast, France's agency for Food Health Safety (ANSES) called for tougher preventive measures in September 2011. In particular, infants and pregnant or nursing mothers were advised to cut bisphenol A from their diets. ANSES warned that even "low doses" of the chemical had a confirmed effect on lab animals and a "suspected" effect on humans, thus requiring action.

This is an example of how complex the act of simply reading a nutrition label can become when further examining the argumentation that underlies each chemical found in the food. In the case of bisphenol A, the ingredient isn't even listed on the nutrition label, as it is a component of the packaging rather than of the food itself. A potential consumer is likely unaware of its existence in the product, even with full examination of a nutrition label. One nation may outlaw a chemical while another allows it. Politics and science often make for strange bedfellows. Changing the composition of U.S. products away from bisphenol A may prove more difficult than for the French industry. One nation may recommend changes, but not another. Over 80 percent of U.S. cosmetics are not allowed on the European market due to health safety risks. The European Union Cosmetics Directive banned 1,328 chemicals from cosmetics in 2003; the U.S. Food and Drug Administration (FDA) has banned or restricted only eleven. On the other hand, many medical procedures and pharmaceutical drugs allowed in Europe are not FDA approved in the United States. Pradaxa, a blood-thinning drug used by over 2 percent of the population, was used in Europe for years before recently becoming available in the United States.

Often it may not be the science but the politics that drive action or lack thereof. Perhaps in the case of bisphenol A, it may be an overreaction by the French authorities because their culture supports a chemical-free diet and criticizes U.S. additives. Only through rigorous testing and mathematical support from those tests—in other words, the tools of science—can the truth of the argument be established.

Consider the notorious Nazi Germany, which restricted asbestos use in the 1930s, citing human health hazards. At the same time, U.S. government agencies continued to support its usage for decades. Asbestos is a wonder material, able to be molded into any shape and size, flame retardant, and an excellent insulator. Unfortunately, that wonder comes along with a human health offset. Even limited respiratory exposure to loose fibers of the substance may lead to lung cancer and other lung-related problems. Asbestos has been linked to lung ailments because the asbestos particles are small enough to enter through the nasal cavities and into the lung, causing irritation. Until the 1980s, U.S. industries used asbestos for insulation and fire proofing in every facet of building. I recall a classmate who would throw his pencil into the asbestos in the ceiling and eat the insulation falling down with the pencil in our junior high school in Queens, New York, in the early 1980s.

EARLY ARGUMENTATION

Philosophers have argued over the truth about phenomena for as long as recorded human history. Aristotle, the Greek philosopher (384–322 BCE) and "father of science," was one of the most important figures in Western philosophy. He wrote on many varied subjects, including physics, astronomy, poetry, rhetoric and argumentation, morality, logic, and science. In fact, Aristotle studied almost every topic in his time, thinking about the existence of science as a means to examine the universe. He was the earliest philosopher to formally study logic and argument. Immanuel Kant credited him in his work, *Critique of Pure Reason*, with laying the foundations for modern deductive inference, a major part of the logic of science used in designing research studies today.

Aristotle used observation and classification of those observations to draw scientific conclusions about the world and universe. He termed the study of the natural world *natural philosophy*, and this was the first organized branch of thinking that made sense of humans' surroundings. Natural philosophy was a precursor to science in that it included the study of both living and nonliving systems. Today, these branches of science would include biology and the natural sciences, as well as physics and chemistry, that are encompassed by the physical sciences. Aristotle's thinking led to generic and abstract inquiries, such as ethics and logic as well as metaphysics and astronomy. He included the full classification of living creatures along a hierarchy of complexity, purported a geocentric model of the universe (with Earth at the center), and studied changes in movement and speed.

While truth existed only in the observable in the ancient and medieval eras, some observations were difficult to explain beyond Aristotle's time. For example, he could not explain why some planets moved backward. A second-century Greek scholar, Ptolemy of Alexandria, offered a solution. He argued that planets moved in smaller circles, each of which turned along a larger circle, to create a backward-appearing effect that made the appearance different from reality. This described the backward variant and was actually an accurate model for planetary motion used today.

Surprisingly, Aristotle's views dominated scientific thought for almost two thousand years. To his credit, he had explained a great deal, and it was not until the sixteenth century that European scholars began to question his viewpoints in an organized way. The ideas of Aristotle and Ptolemy were passed on from generation to generation of scholars in a very conserving intellectual community. The ancient Greeks, through the Byzantine Empire and onto the modern medieval era, explained the universe with Aristotelian observations. Arab scholars and even Chinese thinking about the world were in tandem with an Earth-centered universe, for example. It made sense to them at the time. The discovery of ancient teachings, coupled with the emergence of a new way of thinking that focused on questioning the way things were, led to a shift from Aristotle's views of science to one of skepticism.

KNOWING ABOUT KNOWING

Because Aristotle studied all aspects of knowledge in his time, he is often said to be the last person to know everything there is to know. He used systematic means to observe and study his world. While modern philosophy excludes his methods of empirical study of the natural world, Aristotle included this scientific approach within his natural philosophy discipline. The ways to discover truth and knowledge are divided into two philosophical fields.

The branch of philosophy that studies the nature of knowing and the origins of thought is called *epistemology*. *Episteme* derives from the Greek word for "knowing" and *logos* refers to the "word or the study or science of," together forming the term. "What is knowledge?" "How is it acquired?" and "How do we know what we know?" are the three basic epistemological questions. The search for a belief system and a justification of how knowledge is sought and created is a main goal of epistemological studies. The Scottish philosopher James Frederick Ferrier (1808–1864) first introduced the term, but the study of epistemology dates back to before Aristotle.

Inherent in epistemology is the search for truth or reality behind knowing. It is thus based, in part, on ontology (from the Greek *onto* or "being; that which is," and *logos*), which is the study of the nature of being, existence, or even reality. Ontology is a part of the branch of philosophy known as metaphysics, and it concerns itself with the nature of truth or reality.

In fact, with the explosion of knowledge and the realization that knowledge is indeed uncertain, a modern scientific goal is not to "know everything" but to develop as much understanding as possible, given a certain time and era, within scientific abilities. Instead of merely observing phenomena, modern science seeks to distrust existing knowledge in the constant attempt to better what is known. A surgical procedure or a computer application can always be improved upon to create a more desired product.

Aristotle's teacher, Plato (429–347 BCE), advocated a view of knowing based on such distrust of the known. He worked with dialectics to search for the truth in answering scientific questions. Skepticism and a rejection of appearance were central themes in Plato's views of searching for answers. Plato's view on the world emphasized the abstract while Aristotle only looked at the tangible. To show the contrast, consider the study of vessels within leaves on trees in a forest. Aristotle would classify the vessels according to size and location and give them names. Plato would question the orientation of the vessels to consider the larger picture of how vessels fit and function within the trees. Aristotle's explanation would be concrete, whereas Plato would play with possibilities.

Plato used the dialectic process, which involved one party giving a statement (called the *thesis*) and another arguing another side (*antithesis*). The argumentation would continue in a back-and-forth method, eventually leading to a region of agreement between the two viewpoints called a *synthesis*. The ontological reaching of truth occurred in this way in order to reach scientific conclusions that approximate truth. Georg Wilhelm Friedrich Hegel (1770–1831) is often credited with discovering Hegelian dialectics, but in fact, he only expanded upon Plato's and Kant's views of the process. Nonetheless, dialectics remain a modern form of argumentation used in the sciences to further knowledge. Consider a paper presentation by a scientist on the effects of chlorofluorocarbon (CFC) emissions on global warming. The scientific community is obligated, under the terms of Hegelian dialectics, to debate and debunk the conclusions presented by the presenter of the paper. Often, heated exchanges occur, which can be highly valuable in clarifying scientific ideas. At times political posturing may be present, which can mask information and ideas argued from one side or another, perhaps even compromising the search for scientific truth. This dialectic form of argumentation is, however, a necessary contributor to finding scientific truths. It dates back before Plato to his teacher Socrates.

Socrates (470–399 BCE) was also a Greek philosopher and arguably the founder of the modern scientific method and critical reasoning. He developed his ideas as a reaction against the groupthink of the time. In fact, Athens as a democracy was actually quite dictatorial and tolerated very little dissent. As an outspoken figure, Socrates would roam the streets questioning people and their ideas. He had a following and was well regarded by many people in ancient Greece. In questioning his followers and students, he would refute their answers and use the argumentation dialectic to help bring them to a greater understanding of truth. This type of questioning and answering is now called the *Socratic method*. Socrates never wrote down his ideas, but Plato recorded his many dialogues to preserve his teacher's way of thinking.

Much like dissenters of today, Socrates enraged the powers of the ancient Greek leadership. They asked him to become a conformist and join their side. Their hegemony was threatened by his questioning of a variety of issues at the time, including local wars and economic decisions. They tried Socrates and found him guilty of treason and corrupting young people. He was sentenced to death by drinking a deadly cup of the poison hemlock. Socrates had the opportunity to be saved by exiling himself from Greece, but to this he replied, "Life without enquiry is not worth living." Socrates is best remembered in the scientific community for these words, and they have sparked, in the modern science education movement, a shift toward teaching and assessing inquiry.

PARADIGM SHIFT

The discovery of the philosophers who opposed Aristotle's methods of science gave rise to the scientific revolution. It was a revolution because it was a complete change in thinking—a shift in paradigm from observation to skepticism. Plato and Socrates advocated questioning what was known and what was certain. Plato's emphasis of moving beyond mere appearances to obtain true knowledge is based, in part, on what is termed *Hermetic doctrine*. Hermetic doctrine is based on the writings of Hermes Trismegistus, an ancient Egyptian priest who stated that all matter contains a divine spirit that is unleashed when truth is discovered. Therefore, knowledge exists in nature and through studying the natural world, people learn and a spirit of understanding is common to all phenomena.

Hermetic doctrine moved beyond the obvious and observable and sought to find God in nature and in mathematics. It required a form of questioning and skepticism that pushed outside the limits of the visible natural world to inquire what was inside. Hermetic thought still motivates the advancement of science. During the Middle Ages, alchemy (the study of transforming metals into gold; early chemistry), astrology (the study of how stars affect humans; early astronomy), and magic (the manipulation of human perception; early psychology) witnessed a boom into new ways of questioning based on Hermetic doctrine and neo-Platonic skepticism. An epistemological and ontological shift occurred to begin a paradigm of questioning that is so vital to science today. That was the start of modern science and modern scientific methods, discussed in the next chapter.

LOGIC BEHIND KNOWING

Logic emerged from the study of dialectics, which used opposing viewpoints to continually debunk the other point and to reach some sort of truth at the end of the process. The earlier philosophers used observation and reason to prove points and derive laws of the universe. The prevailing logic at any time may make sense to those in that time and place, but when subject to rigorous testing, many sensible conclusions are debunked. For example, Aristotle believed that Earth was the center of the universe. This was sensible and logical because to the human eye, all stars, the sun, and the moon continually rotate around the Earth in the sky above. This model, termed the *geocentric model of the universe*, dominated scientific thought until the later Middle Ages when Copernicus posited (and was persecuted for it by the Church) the heliocentric model, with the sun at the center. Aristotle also mistakenly believed that heavier objects fall to the ground faster than lighter objects. This remained the prevailing belief until the start of the scientific revolution (1540–1690) in Europe, when Galileo showed that objects of different mass dropped from a tower fall to the ground at the same rate.

While Aristotle was wrong based on his logic in the above examples, his thoughts dominated worldly beliefs for millennia because his observations were profound and his logic was reasonable, given the information of the times. Modern science seeks greater rigor than simple logic and observations, though. In the next chapter, the tools modern scientists use are discussed as a way toward truth.

Logic emerged from the idea that solid premises would lead to clear conclusions by ruling out alternatives. Conclusions based on reasonable foundations would always follow from reasonable premises. Plato had a difficult time developing a clear method for logic and relied mainly on dialectics and skepticism to gain knowledge. Through *reductio ad absurdum* he would break down problems through contradiction and develop them into a reasonable conclusion. Logic works well this way, dealing only with the hypothetical and not with the measurable known. "If there was assuredly no God, what would happen to human behavior?" is a question asked by many logicians but cannot be answered. Modern science seeks measurable answers.

TYPOLOGY OF ARGUMENTATION

Productive scientific research requires a level of argumentative reasoning that asks the right questions to fully consider a phenomenon. The epistemological and ontological foundations of any argument should be explored both by the arguer and the evaluator of the argument. These underpinnings shed significant light on the motivation behind the argument and thus the strength of the conclusions drawn from that argument. This composes a critical evaluation of thinking so vital for successful reasoning and scientific thinking.

In the example used in the previous chapter, a UCLA (University of California at Los Angeles) study, the media reported the role bacteria play in causing pancreatic cancer. Clearly, erroneous conclusions were drawn from a series of faulty argumentative assumptions. While the report showed that *Streptococcus mitis*, the type of oral bacteria that also causes tooth decay, declined in numbers in the oral flora (population of bacteria) of people with pancreatic cancer, there was no logical reason for the media to conclude that there was a link between tooth decay and pancreatic cancer.

This example will be used to show the ontological and epistemological underpinnings that drive scientific thinking about an issue or research question. Scientific thinking and reasoning may be classified into a typology (ranking scale) of argumentation that evaluates the strength of scientific conclusions. Through incorporating my use of argumentative strategy and the works of a variety of educational and psychological researchers, the following typology was developed and placed in the chart shown in figure 2.1 (see page 32). It encompasses the societal, philosophical, and intellectual components encompassing scientific argumentation. A person does not necessarily operate in isolation in considering a scientific question or issue. The mind is complex; it incorporates prior experiences and cultural components as well as reasoning strategies to think about science.

The most basic and simplistic form of argumentation in the typology is Stage 1, *observation* level argumentation, which dates back to the ancient Greek natural philosophers, such as Aristotle, discussed earlier in the chapter. To reasoners at this level, only the *observable* is considered important and measureable. To an extent, simple observation is the basis of science investigation because only that which is sensed can truly be understood. Otherwise, thinking enters the world of nonscience, which often competes and corrupts true science. These nonscience factions will be discussed in detail in chapter 10, "Science Progress and Challenges to Science."

Observational argumentation uses induction to draw conclusions. In this process, the reasoner derives, from facts, some principle that summarizes those pieces of data. Observations and conclusions at this level are acceptable only if they are obtained by the five senses: sight, touch, sound, smell, or taste. Induction, then drawing from or facts to larger conclusions, is the opposite of deduction, which is the process by which facts are gathered from studying the validity of a statement.

In the observation level of reasoning, the abstract or unseen is not considered. For example, Aristotle based his system of organismal classification on simple observations, dividing organisms according to "their manner of life, their actions, and their dispositions using a system of categories involving paired opposites."[4] Observation of animal behavior and plant feeding, for example, led to inductive reasoning and theory building. There is nothing wrong with this kind of thinking, and it often forms the start of any scientific investigation. We all use observation to form opinions about people, places, and events. How often do we judge people by the way they dress, by the neighborhood they live in, or by the house they own? Thus, when thought remains at this level, little deduction occurs and few tests are made. Observation is only the beginning of the scientific process.

Observational argumentation is the lowest type of reasoning. Using the oral bacterial research example and applying only observational argumentation one concludes: "If there is evidence that good hygiene can avoid the terrors of getting pancreatic cancer, then it must be so." A simple relationship is reported about *S. mitis* and tooth decay, and so it is established. In Stage 1 argumentation, observations are truth, and a relationship between tooth decay and pancreatic cancer draws the conclusion. The ontology is that truth is based on our senses.

Naturally, conclusions may not necessarily be the full answer. Stage 2, *authority-determined* argumentation, allows the reasoning to move to a higher power in thinking, and there is appeal to an authority or a governing body for answers to scientific questions. Thus, an individual operating at Stage 2 supports the conclusions of the scientific community. In the example of oral bacteria, because a scientific organization, UCLA, and a media outlet reported the conclusion, the ontology is validated. Majority and popularity of thought prevail at this level, and the conclusion posited is accepted: "Good hygiene can help one avoid the terrors of getting pancreatic cancer." Dangerous thinking occurs at this level because certain powers have significant influence on the public. In our modern society it is largely the media that holds this influential power. Chapter 6, "The Media," will discuss the extensive effects of media on science progress and thinking.

Scientists are not immune from operating within a Stage 2 orientation. Most scientists are part of a larger community, and this prevents them from upsetting group-held beliefs and damaging their own reputations. This level of thought adds nothing to original research because it conforms to a paradigm of science that does not question. It is antithetical to the process of science that starts with a research question about existing knowledge.

In the example of pancreatic cancer and tooth decay, the research question is appropriate in that it seeks to establish a previously unknown relationship between teeth and cancer. However, merely accepting the results because an authority holds the view is the direct opposite way of thinking demanded by science. Science progresses only by breaking through existing thought structures to develop new ideas. If culture, whether large or within the smaller scientific community, prohibits this, it is to the detriment of science.

Stage 3, *belief-based* level of argumentation, incorporates "belief" as a higher level than simple acceptance of societally held conclusions. While belief at this level is not based on evidence, it is placed at a higher stage of reasoning than authority because it at least contests another existing paradigm(s). It may adhere to its own groupthink mentality with other believers, but at least there is a battle of ideas. At this level, there is a conflict between authority and an individual belief system. To illustrate, a belief that there is no truth to oral bacteria's role in cancer at least questions the status quo. Perhaps at this level, the arguer knows of someone, such as a family member, who had excellent hygiene and still developed pancreatic cancer. Neo-Platonic skepticism is incorporated to question the authority and therein lays the seeds of successful scientific reasoning. The epistemological foundations at this stage are still weak because the person does not accept that his or her beliefs may be flawed. Instead, knowledge is absolute, and ontological assumptions are that there is a right versus wrong answer to the issue.

Only in the next stages, 4 and 5, where *evidence-based* argumentative strategies are used, does epistemological and ontological assumption move beyond the certainty that ideas are right and wrong. There may be right or wrong, but at Stages 4 and 5 the truth is uncertain and perhaps more complex than at the earlier stages. There is the consideration of evidence to refute accepted thinking about phenomena or to support contemporary understanding. Use of evidence determines whether it is Stage 4 or 5. The level of questioning and strength of the evidence at Stage 4 is still not sophisticated enough to draw from as many sides of the argument as possible. At Stage 5, the questioning about one's views moves to a gathering of evidence from all of the aspects of the issue. Gathering of evidence is the major factor shifting thinking from the lower three levels of reasoning to this next epistemological stage.

Looking again to the example of the oral bacteria study, upon closer inspection, data are evaluated and evidence is considered and debated to establish the validity of statements made by the media. In both Stages 4 and 5, conclusions by the media are not merely accepted. Stages 4 and 5 are differentiated based on the strength of the evidence used to support a conclusion. Evidence-based decision making is the key factor in moving an argument from one stage to the next in classification. Upon further evaluation of the data in our example, the report is obviously erroneous in its conclusion. This level realizes that hygiene does not prevent cancer because the media's conclusions do not match the numbers given. People with pancreatic cancer actually had fewer oral *S. mitis* cells in the reported data, so effective tooth brushing would have the opposite effect—increasing pancreatic cancer susceptibility.

So why does the report make such an erroneous and outrageous claim? It sells newspapers and airtime. All individuals like to feel empowered so that they can somehow avoid dreadful diseases and in some way elude fate by their own hands; in this case, avoiding pancreatic cancer by brushing. The report is positive, interesting, "headline grabbing," and makes sense to the public. Media outlets grabbed at the chance to report the study erroneously. However, the media's reporting is an example of bad science in every way. It is contrary to the evidence-based reasoning required at the 4 and 5 argumentative stages.

In the final and highest stages, 6 and 7, the *probability-based* argumentative strategies, there is a complex and varied use of criteria to evaluate the strength of an argument or arguments. The strength of the evidence, the strength of the alternative evidence, and the risks of making an error in accepting each of the conclusions are evaluated to determine the truth of a conclusion. Epistemologically advanced stages view knowledge as uncertain and able to be changed with new or different perspectives and evidence. To illustrate, Stages 6 and 7 use evidence in ways to bring forth a strong case to find the ontological truth of a particular premise. In short, the argument is strong only when a rigorous evaluation of the strength of the information presented is conducted. It requires the many mathematical and process tools that scientists use in their methods of analysis.

The conclusions in the oral bacteria example are evaluated based on methodology, experimental design, or the mathematics backing the data. In fact, upon further investigation at Stages 6 and 7, an individual will move beyond the news media to other more rigorous science journals. In the example given, the journal article shows that only ten subjects were tested in the study and no real claims could reasonably be made beyond the preliminary. A sample size of ten is very small and lacks the statistical power to support generalizations made by such a study. The person at these higher stages understands the danger of a media that calls for national testing of people's mouths for pancreatic cancer based on the study.

A good argumentative strategy at the higher stages would evaluate the best evidence possible but also explore the strength of the "counterarguments." Counterarguments are defined as arguments against the original premise that work to discredit that premise. This information can be used to refute or support a hypothesis, depending on how it is presented and how it is placed within an argument. The highest level, Stage 7, would also evaluate the possibility of error in each piece of data supporting the arguments. Such an argument would delineate the mathematics, research design strength and validity of the measures, study works cited about the topic, and evaluate the opposing evidence of an argument to determine errors on both sides.

There are inherent probabilities of error in every research study. First, mere relationships in the first argument do not imply causation, a major mistake often made to exaggerate research results. Perhaps there are other intervening variables. In the example study, other variables may exist in addition to tooth-brushing habits and pancreatic cancer rates. The absence of bacteria could be a sign that health is deteriorating and changes are thus a result of pancreatic disease.

Unfortunately, there is also pressure on the scientific community to produce positive results from investigations. The *Journal of Medical Ethics* reported a sevenfold increase in retractions in research related just to errors in the period between 2004 and 2009.[5] Clearly, there is a need to question research on many levels, including the errors in their argumentative strategies.

In order to determine the strength of any evidence in a scientific argument: 1) There should be an assessment of the relationships between the evidence and the scientific conclusion in question. Does the conclusion match the data? 2) There is a raw tally of the number of facts or issues addressed versus the total number of possible ones to address in an argument. How inclusive is the argument in terms of how it covered the topic? For example, does each oral bacteria argument include all available data to defend the ideas, or did some get left out? 3) Does the evidence in support of one's argument counteract the counterevidence? Being able to argue the other side's perspective is important in any argumentation strategy. Although oral bacteria have been implicated in many diseases, ranging from heart disease to inflammations, debunking links between oral pathology and cancers weakens the argument shown in our example. It is difficult to judge the true strength of an argument beyond these points.

The chart in figure 2.1 on the following page summarizes the stages of argumentative reasoning presented, dividing them based on their shared and divergent assumptions, ontologically and epistemologically.

EPISTEMOLOGICAL PLASTICITY

Is an individual locked into a particular stage of reasoning, or is development beyond that stage possible? A variety of philosophers and educational researchers explored this notion, with varying answers. Jean Piaget (1896–1980), a famous educational researcher studying childhood thought development, argued that reasoning ability was inherently fixed at certain ages for individuals. He established the branch of epistemological study termed *genetic epistemology*, which claims that thought development is set based on age and individual genetics. Without a hope for change beyond certain intellectual milestones, in his model, nature is the determinant of reasoning. It is possible that some people develop intellectually and others do not.[6]

Another set of educational researchers, as far back as Lev Vygotsky (1896–1934), contend that humans are not locked into Piagetian reasoning levels. Vygotsky studied children's play and speech patterns and found that there is a level of appropriate instruction, based on a student's ability level, to help them to develop their reasoning patterns. Given the right scaffolding, Vygotsky argued, the child improved in scientific thinking skills. This level of instruction is known as the *zone of proximal development*.[7]

MOTIVATION AND REASONING

A group of schoolchildren play each afternoon in a sandbox, enjoying each other's company and activities for hours. Then one day, the teacher tells each of them that they will earn a dollar a day just for playing in the sandbox. The children are ecstatic because they are now getting paid for something they already love to do. There is only one catch—they must show up on time and play for at least one hour to earn that dollar. At first, it is money and play and every child is happy. Then, each child quickly grows to hate playing in the sandbox. Because they are being paid and it has

Stage No.	Stage Name	Descriptor	Epistemology
1	Observation	What is observed is truth	Certainty
2	Authority	Popularity or authority determines truth	Certainty
3	Belief	Conflict between authority and individual belief system	Unsupported uncertainty vs. certainty
4	Evidence	Data and valid facts relating to conclusions	Approaching uncertainty
5	Evidence	Stronger data and valid facts relating to conclusions	Uncertainty
6	Probability of truth	Evaluation of evidence based on: -strength of evidence -strength of alternate evidence -risks of error in accepting evidence	Uncertainty
7	Probability of truth	Stronger evaluation of evidence: based on: -strength of evidence -strength of alternate evidence -risks of error in accepting evidence	Uncertainty

Conclusions are Plausible Points of View:

Evidence strength depends on:

1) relation to question

2) coherence (data matching conclusion)

3) counteracting alternative hypothesis

4) number and quality of points addressed total possible points

5) risks of error

Figure 2.1.   Typology of argumentation: Justification of beliefs.

now become a job, they do not want to do it. Soon, the system breaks down—within weeks, in fact—and no child plays in the sandbox any longer.

This story is based on real data from a variety of research studies on motivational behavior. The children are perfectly normal, but external factors (for example, teacher interference, required work) inhibit their motivation. The addition of a reward system works in opposition to the intended effects. The story shows how human behavior is not always simple to understand, and in fact, our complexity cannot be overstated.

Similarly, the human ability to learn is based on motivation. In the case of improving reasoning, many studies show that motivation by the learner is the single most important factor in determining the development of one's intellectual capacity and reasoning skills. While Piaget may be correct in showing the difficulty with which cognitive changes may occur, Vygotsky's thinking is closer to predicting the outcomes for most children and even adults. The power of motivation (and demotivation), as seen in the sandbox story, is advocated as a means to improving reasoning in schools and in society. A motivation to be better thinkers in science is a major theme of this book and is addressed in later chapters as a goal to improving societal and scientific progress.

An individual may be at one level of reasoning on a certain issue and, at the same time, be at another level on an unrelated topic. A climate change opponent may not be able to reasonably consider counterevidence supporting climate change. The person is at a lower level of thought in this scenario. However, she or he may be able to critically evaluate whether or not to do a health care procedure on a personal level. In this example, the environmental topic may change one's level of reasoning because the topic has personal meaning. That may interfere with a person's ability to reach higher levels of thought. The medical procedure may not be laden with personal commitment, so it may be considered without impingement.

As stated earlier in the chapter, scientific thinking occurs within a cultural and intellectual edifice, with complex and multifaceted aspects contributing to an individual's reasoning patterns. Unfortunately, according to several researchers (e.g., Hofer and Pintrich, 1997; King and Kitchener, 1994), most adults lack the advanced reasoning patterns described earlier that are fundamental to scientific literacy.[8] In the absence of such cognitive capacity, there is a need to either develop those skills or to guide the media and educational outlets to properly present scientific information. Unfortunately, the latter is impracticable due to a host of competing agents, so a motivation to better public science thinking needs to be established.

PHILOSOPHY OF SCIENCE

As is established by the study of reasoning and reasoning development, the ability to argue points and analyze problems in science is central to a philosophy of how science works to change society. While philosophical movements are complex and often changing, there are several scientific philosophy branches that explore the nature of knowing and creating knowledge. These epistemological outlooks each purport a certain way of thinking but blend with each other and change in practice. These shifts will guide the way science is done. Ways of understanding truth through science are the basis of all four philosophical perspectives that will be presented. These differences are likely to guide scientific thought in the twenty-first century.

There are continued questions in society that demand a scientific approach, based on reasoned philosophy, to give credible answers. Consider the potato, a healthy and nutritious vegetable, which has nourished humans since the Middle Ages. A host of news agencies reported research results on the effects of potatoes on public health, and they were very negative.

To illustrate, a *Wall Street Journal* article reported the study in an article titled, "You Say Potato, Scale Says Uh-Oh," implicating potatoes as a major cause of weight gain, and cited in its front cover photo that "boiled, baked, or mashed potatoes correlate with a .57 pound weight gain" per year for people. This implied that all potatoes, even the nonfried ones, are bad for human health and society. Surprisingly, when investigating further beyond the article's facts, one finds that the nutritional information on potatoes is actually very favorable.

Consider that a boiled potato, cooked in its skin without salt and butter, has only sixty-eight calories, is very low in saturated fat, cholesterol, and salt, and is also a good source of vitamin B6, potassium, and vitamin C. Is it the extra butter and cream in mashed potatoes or the sour cream and butter on baked and boiled potatoes that likely lead to the weight gain cited?

The article omits these extraneous variables co-occurring with the potato-weight-gain link. So what is the motivation? Is it merely an oversight? Is it part of a larger groupthink mentality and a social movement to eliminate carbohydrates from diets? Did ulterior motives to satiate that movement give the newspapers an audience for their anticarb campaign?

It is not a lone error or the fault of biased reporter, because many news outlets reported the same story in November 2011. Instead, it is part of a larger movement. As will be discussed in later chapters, the media's reporting of scientific information is often truncated and the public is misled about scientific research results. Science combats these nonscience and/or biased reports through a philosophical approach to obtain truth. None of the philosophies are wrong; they simply work within different means to create a common goal—to understand truth behind phenomena.

POSITIVISM

The traditional view of science, since the scientific revolution, holds that a process exists with certain specified steps to reach a conclusion. That series of steps is called the *scientific method*. It is based on supporting or failing to support a hypothesis, or an educated and informed statement. The branch of philosophy based on the scientific method is called *positivism*. Positivism bases knowledge solely on data gained from sense experience. Positivism suggests a hypothesis for a research question and uses the tools of science—through various investigations—to find answers.

Positivists are rather certain in their answers to questions. The conclusions formed from positivism are rather definite, with the power of the scientific method behind them. Positivism holds that larger, deeper-meaning questions cannot be measured and therefore cannot be answered. "Why?" and "What will be?" are not able to be sensed through positivist methods and thus are not a part of scientific philosophy. These are moral and/or ethical questions, which should not be tested. In the case of the potato study, motivation behind the results is irrelevant. Numbers and the results they point to are the only important ideas.

MODERN BRANCHING OF THE PHILOSOPHY OF SCIENCE

The goal of all of the philosophers of science, from the ancient Greeks (Aristotle, Socrates, and Plato) to the European revolutionists (Galileo, Copernicus, and Johannes Kepler), was to find truth. Their methods may differ, but their goals are the same.

Postpositivism (also called postempiricism) is a scientific philosophy that views scientific knowledge as always able to be changed and not based on any real certainty. It criticizes positivism in that it never accepts a hypothesis. Instead, according to the postpositivists, there is no proof in science—only disproof of alternate working hypotheses. Postpositivists argue that a hypothesis is never proven because something could always be found to discredit it. Science progresses through a process of elimination of alternate working ideas. This procedure is termed *falsification*, and it is the basis for modern medicine. A symptom's cause is never fully known until a series of "ruling out" tests and procedures are performed to eliminate possible causes. Sometimes there are several symptoms for a cause and several causes for a symptom.

Karl Popper, who wrote many treatises on the philosophy of postpositivist science, is its founder. Take the example of a medical problem such as a pain in the kidney. Do we know if it is a kidney stone or an infection or even cancer? Falsification leads to uncertainty. Understandably, patients are averse to this uncertainty. Therefore, medical doctors who use the postpositivist paradigm to falsify their diagnoses frustrate people.

Postpositivism deemphasizes the importance of an absolute truth in the universe. It reasons at a higher level of argumentation in the typology than positivism in that it recognizes the uncertainty of knowledge. In the case of the potato study, the numbers may clearly show that there is bias in attacking the potato, but the postpositivist does not know for certain that the potato really is good for you. Instead, there could be some unknown substance within the potato that actually causes weight gain. Only through more and more tests and investigations would the postpositivist actually more certainly know about the potato's influence on human weight. In essence, this philosophy operates in direct contradiction to modern media reporting. The media seeks simple solutions and instant results and conclusions. Postpositivism takes a great deal of time and effort even to consider support for a hypothesis.

Two newer philosophical branches of science, relativism and realism, emerged in ontological contrast to both positivism and postpositivism. Relativism contends that all knowledge is true only insofar as it is compared with something else and exists within cultural confines. Given the potato study as an example, societal influences drive the study and affect interpretation of the data. Thus, society plays a role (whether right or wrong) in developing conclusions and action based on the science.

Truth for the relativist is determined not as certain and real, but as related to how the data appear surrounded by other studies or within the society. In relativism, the role of society in scientific progress is emphasized. It argues that society creates various truths and changes only as the community's views change. The potato itself may be nutritious, but within our society it is not viewed as so. Our food preparation methods (adding butter and cream or french frying the potato) are likely responsible for this unhealthy view. Relativists recognize the societal opinions of potato health as an important influence. Instead of a move toward some objective truth, as seen in positivism and postpositivism, relativism sees shifts in societal paradigms as the impetus for changes in science beliefs. As a result, this potato study and its method of presentation hold the potential to change the diet of the U.S. population.

Relativism has serious drawbacks for the scientific community. It downplays that objective reality may exist! It argues that societal constructs and not phenomena make reality. Obviously, a pure version of relativism is unacceptable to scientific thinking because it fails to take into account the reality that some objective and nonhuman truth is out there.

Another branch of modern philosophy is scientific realism. Realism is defined as a view suggesting that the world described by science and its methodology is the real world. A realist philosophy is concerned with science's application to human concerns. Realism looks at the benefits of science and its applications to the real world but does not emphasize social importance, as relativism does.[9]

Realism is, to some extent, a combination of both postpositivist empiricism and relativism. Its ontology is based on observable data that can be falsified, as in empiricism. Its link to society and relativism relates to its societal application. Realism supports a practical way of solving problems in the world. In the case of potatoes in U.S. diets, the realist argues that the research needs to be reflective of the truth. It also contends that society needs to implement those results; otherwise the study is useless. The goal of a study to find real solutions is a goal of realism. However, similar to other modern philosophies, realism accepts that there is some knowledge unknown, and in the case of the potato, further research could impact dietary recommendations.

The realist accepts that science cannot know everything but that its methodology seeks to find and apply real results. "Do chiropractic methods and acupuncture really work?" "Is there a real biochemical challenge to evolution?" "Will we cure cancer, or is it just hype?" These are questions the realist attempts to answer within the confines of society. Realism is practical in its understanding that research can be manipulated or represented in different ways.

In order to evaluate science with an open mind, the philosophical origins of research should be understood. This chapter delineated the many facets of evaluating the beliefs behind science and knowing. How results of a study are used depends on the philosophical perspective of both the researcher and society interpreting that information. The underlying philosophy of any scientific investigation results in different methods by which tests are performed, biases of the researcher, societal pressures on scientists, and research design. The perspective of the reader is vital in judging knowledge claims and in interpreting scientific discovery. The next chapter investigates the many tools scientists use to explore unanswered phenomena.

3

Tools Scientists Use

In August 2007, a twelve-year-old boy was admitted to the hospital in central Texas after attending a summer camp in the weeks before his illness. His mother reported that her son had been disoriented and lethargic, with a six-day history of fever. While at the camp, the boy visited the nurse's office several times complaining of "not feeling well" after participating in recreational water activities in a lake cove. After the hospital admittance, several incorrect diagnoses were made, including meningitis, pneumonia, and bacteremia. The boy's cerebrospinal fluid (CSF) showed an opaque appearance, bloody in color, and a high white blood cell count of 1,750 cells/mm3 (normal: 0 to 5 cells/mm3), red blood cell count of 30,750 cells/mm3 (normal: 0 cells/mm3), a glucose level of 92 mg/dL (normal: 40 to 70 mg/dL), and a protein level of 88 mg/dL (normal: 15 to 45 mg/dL). Amoebas were then identified in his CSF and later identified as *N. fowleri*. Despite aggressive treatment with amphotericin B, rifampin, and azithromycin, the boy died five days after being admitted. The average temperature of the lake he was swimming in at the camp that summer was 84.4 degrees Fahrenheit (29.1 degrees Celsius), certainly warm enough to support the life cycle of amoeba parasites.[1]

The story above is an actual report by the Texas health authorities chronicling the events surrounding the death of a Texas boy. He became infected with *Naegleria fowleri*, a freshwater single-celled protist that enters the brain and aggressively destroys tissue. This is a rare illness known as amebic meningoencephalitis. Pond water must have entered the patient's nasal passages, through which the amoeba was then able to pass into an area of the skull between the sinuses and the brain linings known as the cribriform plate. The parasite then lodges itself in the olfactory bulbs, which are large frontal regions of the brain right above the nasal passages that sense smells, as shown in figure 3.1. Amoebas divide here, causing necrotic (dead tissue) lesions that affect the sense of smell first. Often, symptoms include a loss of smell or a sense that something is burning or rotting. Amoebas then rapidly advance into the brain,

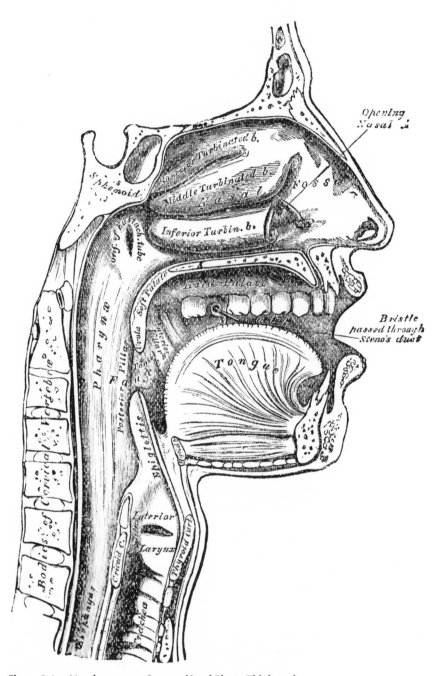

Figure 3.1. Nasal passages. Source: iStockPhoto/Thinkstock.

dividing and damaging many areas until they attack the brain stem, leading to the victim's death.

The course of the disease generally takes between three and twelve days. Its survival rate, with aggressive medical treatment, is only 3 percent. A five-year-old and sixty-one-year-old are known survivors. A main reason for the high mortality rate is that the disease is often misdiagnosed (as in the case for the above victim) because it resembles bacterial meningitis. Symptoms of amebic meningoencephalitis include headache, fever, nausea, vomiting, and a stiff neck in the first few days of infection. After about a week, symptoms progress to confusion (as brain tissue is destroyed), lack of attention to the environment, loss of balance, seizures, hallucinations, and death.

N. fowleri is found in hot springs, warm freshwaters, and according to a recent CDC (Centers for Disease Control and Prevention) report, well water and municipal water systems. The latter is concerning because of the amoeba's ability to get into everyone's water supply. Ingestion or drinking of infected water is harmless; it is only potentially harmful when the infected water gets too close to the nasal passageways.

Two cases of infected individuals who used neti pots have been reported recently in the United States, calling into question the use of the devices. The neti pot is used to irrigate (clean) nasal systems. Water is poured from the pot through one nostril into and flowing out of the other. It is an ancient Eastern technique, now being increasingly recommended by physicians for allergy, cold, and sinusitis treatment. *N. fowleri* was found in the home water supply of both victims. Apparently, the amoeba is able to move from the water in the neti pot through nasal passages to the cribriform plate and up to the brain to cause infection. Obviously, if sterile, distilled water is used with the neti pots (as is advised in their instructions), the possibility of this rare danger can be avoided. It may be much easier to simply use tap water, but it is not in a person's best interest.

Amebic meningoencephalitis is a complex illness that manifests differently in each individual. Science shows us that the amoeba can penetrate the nasopharyngeal mucosal layers of the nose and migrate into the olfactory nerve cells of the brain through the cribriform plate region. From 1995 through 2004, *N. fowleri* killed only twenty-three people in the United States. This is clearly an extremely rare illness, with only 160 cases reported since the amoeba was identified in the early 1960s, according to the U.S. Centers for Disease Control and Prevention. This disease remains a poorly understood medical mystery because of so many unanswered questions. What makes some people susceptible while others go unharmed when swimming in waters laden with the pathogen? Why is the disease such a rare occurrence? How does the amoeba push its way through all of the protective membranes?

SCIENCE IS COMPLEX

Amebic meningoencephalitis shows the complexity with which scientific phenomena are approached in scientists' study of the unknown. The unanswered questions in the disease course is where science begins and is what sparks interest in studying science. Most people have heard about the scientific method as a means to investigate scientific questions. While there are structured steps for conducting scientific investigations, research is clearly more complex. Scientific research involves many factors, including revisiting stage after stage of study. In the meningoencephalitis case, there is a clear need for the use of the reasoning strategies described in the previous chapter to successfully conduct scientific research.

Specified steps of the scientific method are given in figure 3.2. The scientific method is a basic guide to solving scientific problems. It shows the general way in which science progresses. The start to a science investigation is an inductive analysis of a problem. It is the excitement stage of research. In the example given, the shock or compassion for the victims of amebic meningoencephalitis drives the researcher to study the problem of how the disease is manifested. It begins with a simple observation by the researcher to gathering information about the topic. Perhaps through a critical review of existing knowledge in databases and libraries, evidence is gathered to develop a hypothesis. A hypothesis, then, is more than a simple "educated guess." It is a foundation on which a passion for the topic develops.

The hypothesis is a testable question that often stems from research of a body of preexisting information. A hypothesis must be empirically testable, logical, and clear. While a hypothesis is tested, there is some degree of guesswork based on prior knowledge. However, it is more than a simple guess. Consider two final characteristics of hypotheses: furthering science and studying natural phenomena. There are a host of competing agents to science that will be discussed in later chapters. Transgressions against these aspects of hypotheses are the basis for pseudoscience fallacies.

Sometimes hypotheses appear ridiculous, almost pseudoscience-like, and are rejected by the dominant scientific community. Such a faulty rejection of a fellow scientist's hypothesis occurred in 1985. I recall a research presentation by Barry Marshall, who claimed that a bacteria, *Helicobacter pylori* (and not stress and diet) was a main cause of peptic (stomach) ulcers. Marshall's hypothesis ran counter to everything the science community knew about peptic ulcers. Because it overturned the accepted way of thinking, Marshall was met with ridicule and insult in his presentation of data and ideas. The prevailing science of the day pointed to stress and diet as triggers for peptic ulcers. Existing research at the time acknowledged that an ulcer might contain bacteria insofar as it was merely an environment conducive for bacterial growth.

The reluctance of his colleagues to accept the idea that *H. pylori* caused ulcers provoked Marshall to the extreme action of drinking a vial of *H. pylori* himself. Marshall recalled in a recent interview:

> Those were frustrating times for me. Most of the experts believed that the presence of *H. pylori* in those who turned up with ulcer problems was just a coincidence. I planned to give myself an ulcer, and then treat myself, to prove that *H. pylori* can be a pathogen in normal people. I thought about it for a few weeks, then decided to just do it. Luckily, I only developed a temporary infection.[2]

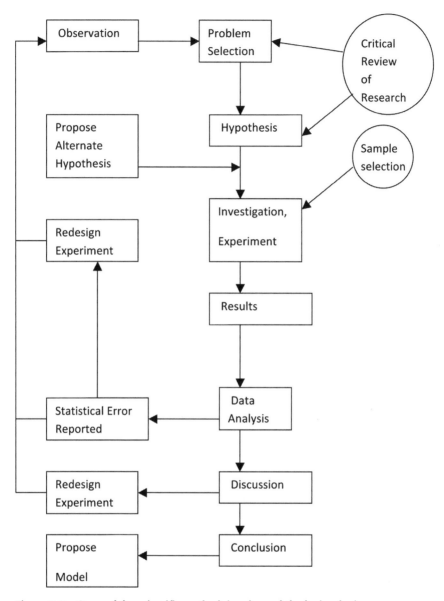

Figure 3.2.   Steps of the scientific method: inquiry and the logic of science.

To the surprise of the scientific community, he formed a number of ulcerations in his stomach. This showed that his hunch was right—that bacteria caused gastric ulceration. Whether or not his methods are advised or ethical is another matter.

RESEARCH DESIGN MATTERS

Science research is traditionally portrayed as conducting an experiment to yield answers; however, a scientific investigation can take many forms: an experiment, naturalistic observation, model formulation, and nonexperimental research design. For the most part, traditional science relies upon experimental designs to elicit numbers and data that may be used to make generalizable conclusions about natural phenomena. In the example of whether or not amebic meningoencephalitis can result from usage of neti pots, the data linking these variables would need to be looked at statistically. There would need to be controlled groups using neti pots with tap water versus distilled water. The independent variable (controlled by the researcher) would be the water type used, and the dependent variable (outcome of the study) would be the level of pathology or illness reported. The results would be statistically analyzed (numbers give power to results in experiments). Conclusions would then be drawn based on those numbers.

An experiment would demonstrate whether or not the type of water used in the neti pot actually has an effect on the contraction of amebic meningoencephalitis. As for the news reports linking the variables, there were only two cases of people in Louisiana cited to have contracted the illness from the use of tap water. Statistically, two cases are not significant to support a scientifically sound link; further experimentation or data would be required.

Ethical considerations, however, prevent the design suggested above for further experimentation of the linkage. Should people be placed at such possible danger to rinse their nasal passageways with tap water when they could be at risk of this nearly fatal illness? What society would be willing to risk the lives of a few to save many? In history, it was commonly seen (for example, studies of syphilis on African American airmen in the Tuskegee experiment and cancer drugs on humans done today), but is it morally right or wrong?

In addition, what kinds of differences (other variables) exist between people that would conflate the results? In other words, individuals placed within a study group receiving the tap water may have significant differences in their physiology (the way their body functions) than those who contracted amebic meningoencephalitis. It is possible that such physiological variables contribute more to the contraction of the illness than the tested variable itself (type of water used). These variables outside of the experiment that may influence results are termed *extraneous variables*. Extraneous variables often weaken the results of an experiment. Thus, special care is taken to remove outside factors that could affect the results of a study. Consider our use of data in the first chapter showing how smoking does not cause lung cancer. Sometimes numbers can be misused, purposefully or not, to come to incorrect conclusions.

There are many forms of what is termed *quantitative research*. *Quantitative analysis* is defined as the reporting and use of numerical data and methodology. As described above, it is the more traditional scientific analysis system. Quantitative research produces data results in the form of numbers and patterns from which conclusions are then drawn. Large numbers of subjects and controls of their conditions are used to give mathematical power to the research. To illustrate, consider observing the effects of a drug on rheumatic hand joints. Large numbers of individuals would need to be sampled to allow for adequate statistical analyses. Statistics is the study of the collection, organization, analysis, and interpretation of data. The numbers need to quantify symptom improvement for patients and side effects of the drug to determine whether the drug should be available for use.

Consider another, more practical research issue: most adults in the United States bathe (or shower) every day. Some people, however, bathe less frequently or irregularly, claiming that less frequent bathing is better for their skin and saves water. In addition, many of those individuals who bathe less frequently shun the use of deodorants and other cosmetics as well. Reasons given to cleanse less and smell more like oneself include the following:

- Compared with past, more agrarian lifestyles, modern society has more passive lifestyles and people get less dirty.
- Bathing too frequently reduces the population of protective staphylococcus-type beneficial bacteria on the skin, making the individual more susceptible to skin infections including MRSA (methicillin resistant *Staphylococcus aureus*).
- Bathing too frequently with soaps and shampoos removes protective oils from the skin, causing dry skin, cracking, and a variety of undesirable skin conditions.
- Deodorants seal the pores in the skin, leading to unnatural and unhealthy skin conditions.
- Bathing is not ecofriendly and wastes valuable water and the energy needed to warm it.
- The commercial producers of soaps, shampoos, hair conditioners, deodorants, and other cosmetic products promote excessive bathing in order to promote higher consumption of their products and to make money irrespective of human and environmental health.

A number of interesting research questions arise from the topic of bathing rates. What is the optimum frequency of bathing for adults in our modern society? Are the suspected health effects attributed to bathing too frequently real or imagined? Are deodorants, shampoos, and other items a daily necessity in today's world where adults encounter less dirt and perspire less, or have we been successfully manipulated by corporate advertising? A research study would need to quantify the results of different groups of people who bathe at different intervals by measuring their hygiene, overall health, skin condition, and more. Quantitative analysis (measurable results) separates science from the many forms of pseudoscience that threaten empirical results. These pseudosciences are easy to identify because they lack the research design and mathematical analysis required to make conclusions. These will be further discussed and debunked in chapter 7, "Pseudoscience."

It is unduly conventional to classify scientific research into either an experiment or something nonvalid. It is true that the experimental design, with controlled variables and sets of data that can be mathematically analyzed, comprise the most generalizable results. But science has relied upon so many different methods of obtaining information that these need to be addressed and respected.

STATISTICAL TOOLS

> Philosophy is written in this grand book, the universe . . . It is written in the language of mathematics, and its characters are triangles, circles, and other geometric figures without which it is humanly impossible to understand a single word of it.
>
> —Galileo Galilei[3]

Without mathematical analyses, scientific research is powerless, and science blurs with the pseudosciences. As expressed in the quote above by Galileo, an Italian natural philosopher of the 1600s, mathematics interprets research results, and it is the language of science. When a research study is presented or published, it must prove to the science community that its results have mathematical backing; it is not enough to state that there is a difference in two groups studied. It must be shown using numbers. "How much?" and "Is it really enough of a difference?" are the next two questions that should be asked. In the study of tap water links to amebic meningoencephalitis, it would be irresponsible of the research community to make recommendations based on data from studies without enough subjects or with poor statistical results. In the example, only two people died of the illness using neti pots, so mathematically, thus far, this would indicate that there is little danger in neti pot use.

When evaluating a scientific study, it is vitally important to examine the numbers presented (more than the results described by the authors) to draw your own conclusions. When no numbers are given, the next question should be "What are the authors hiding?" Becoming critical of research presented is so important in a world that is becoming inundated with instant, truncated information.

Media outlets often claim that a proper presentation of the full mathematics of scientific studies is impossible due to limited space. This presents unique challenges for the public because it is not given the opportunity to evaluate data. The public, instead, is presented is only the conclusions made by scientists as interpreted by the media. First, I find this unethical; by merely showing conclusions, media reporting agencies weaken knowledge and contribute to science misuse. The neti pot example shows how people could be misled to stop using a perfectly effective device through fear of statistically insignificant links with amebic meningoencephalitis.

Second, I do not believe the media's excuses of space limits for science. Sports pages compose almost half of local papers and whole sections of most national newspapers. It is abhorrent that space is not given to mathematical backing for medical research and health advice (for example, heart surgery results), which impacts people's survival, but sports results are allocated entire sections. It is about priorities, and science is not on top.

Third, omitting such information is dangerous for public health and is deceptive. Giving only part of the research—the conclusions—misleads people. Various stakeholders may use or misuse research (sometimes going so far as to fund it) to propel their own causes. Many times there is a power grab based on partial information provided by media sources. For example, a company producing nasal decongestant products (a market competitor for neti pots) might reap financial benefits by encouraging the media to report incomplete information regarding illness cases linked to neti pots. This leads one to question the ulterior motives behind research presentations by the media. Could pharmaceutical companies be pushing the link between research on neti pots and illness? Is there an attempt to sell more drugs for sinus and allergy symptoms by removing the neti pot competitor? Conspiratorial questions such as these might not need to be asked if the presentation of the research was more complete.

Fourth and most important, science literacy requires a certain amount of knowledge of mathematics to understand and interpret the results of research. Research reported by the media or in scientific papers should be evaluated based on several statistical measures. In every research study reported, valid tests should be performed on the data to show what is termed a *significant* result. An analysis of distributions is performed to determine the strength of the results of the study. Simply measuring the average or median scores only shows the central tendency of a series of data points; it does not show its distribution or dispersion of numbers around the center. Simply stated, the more the scatter of data points around the mean, the weaker the pattern and the weaker the effect of the independent variable that is tested. A narrow distribution shows a stronger effect of the independent variable. Please see the data dispersion graphs that follow in figure 3.3.

Graph A is narrower, so the results are more streamlined around the mean and the results of the study would be stronger. An article reader would see that the study with these numbers has power, and its results are more respected. The spread around the graph is known as the *variance*.

Variance and standard deviation are both measures of dispersion. The variance is calculated by subtracting the mean from each score (x) in a group tested, adding them together and squaring these differences, then dividing by the total number of measurements (n) minus one. Simply dividing by the number of measurements tends to underestimate the variance value.

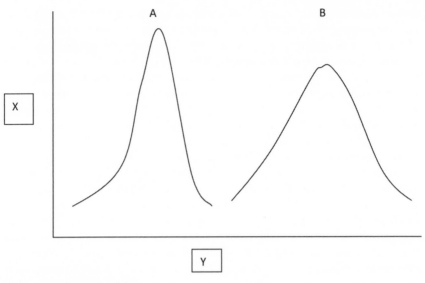

Figure 3.3.   Strength of data.

$$\text{Variance} = \frac{\sum \left(x - \overline{x}\right)^2}{n-1}$$

Standard deviation is calculated by taking the square root of the variance. The reason for using the standard deviation in statistics and not the variance is because it retains units of measure (height in inches, days of rainfall, and more). Calculating the square root compensates for squaring the measurements in the variance calculations. The formula for calculating standard deviation is shown below.

$$\text{Standard Deviation} = \sqrt{\frac{\sum \left(x - \overline{x}\right)^2}{n-1}}$$

Knowledge of how to work through these equations statistically is not necessary for interpreting research results for the scientifically literate. However, understanding that these are the bases for which statistical designs and tests rest upon is important.

A powerful statistical test, called the ANOVA (Analysis of Variance), was developed in the 1950s to show the statistical difference between different groups of subjects' data examined in a study. The ANOVA is a test that identifies and isolates the sources of variation during hypothesis testing.[4] The ANOVA reduces extraneous (unimportant) variables' effects and compares the means of three or more groups in an experimental design. Very briefly, it is a process of isolating two sources of variation within data sets: one source showing the effects of an independent variable (which is controlled by the researcher in the study) combined with unavoidable effects of experimental errors, and the other source reflecting the effects of experimental errors alone. By subtracting the two numbers, the effects of the independent variable (which is the important one being tested) are pinpointed. That is the point of an experiment and data analysis, and that is what an ANOVA accomplishes.

The ANOVA's statistical methods use means and standard deviations to give what is termed an *F-ratio*. The *F*-ratio shows a number that represents the amount of difference in the groups attributed to the independent variable. Minimizing the individual differences between subjects helps isolate the effects of the independent variable. Consider a study testing the effect of humans' eating almonds versus no almonds on their HDL (high-density lipoprotein) cholesterol levels. Group differences could be attributed to outside variables (those other than almond intake) that could have intervened. One group could have cheated and reported fewer almonds eaten or the groups could even be physiologically different from each other, conflating the results.

In general, humans are difficult to study because of the extensive differences between individuals often unaccounted for in experimental design. Thus, disciplines studying human behaviors are often termed "soft" or social sciences. Psychology, anthropology, sociology, and economics are common examples. They are not weaker, of less importance, or of less rigor, but they do lack statistical power behind their results and conclusions.

When *F*-ratios are reported, a probability of error is given along with the number (for example, $F = 2.06$, $p > 0.05$), with p representing the probability. Probability is portrayed statistically as a *significance level*, which is assigned to each study or test by the scientific community. The significance level is defined as the percentage of chance that the results shown are wrong. Recent research has debated the utility of the PSA (prostate-specific antigen) test, which looks at antigens (proteins) for cancer of the prostate in the blood. The test has a significance level of 0.2, which means that its results are wrong 20 percent of the time. This kind of error is common in medical tests and diagnoses. In the case of the PSA test, recent evidence shows that false positives have led to unnecessary biopsies and treatment. A recent recommendation by the American Medical Association to limit the use of the PSA test concludes that patients would have been better off without intervention. But statistics do not always tell the whole story. Whether a person should be tested for PSA, especially after age seventy-five, will be further examined in chapter 8, "Debunking Science Myths: Separating Fact from Fluff."

The medical community often assigns odds to a prognosis: a fifty-fifty chance of recovery from an accident; for pancreatic cancer, a 5 percent survival. There are always unique cases, with placement of a tumor in an odd location or a different physiology, to make certainty impossible. Statistics in science are only a beacon but play an important role in guiding conclusions. Mathematical analysis is an inherent characteristic of science—it shows that all that is certain is not certain!

In science, there are many approaches to limit the innate error in every investigation or test. In the example of prostate cancer diagnoses, the doctor physically checks the prostate for shape and size. Generally, a smoother prostate indicates less chance of cancer compared to rougher. This "digital" exam is used in conjunction with the PSA test and symptoms to further reduce the chances of error.

THE ELUSIVE CORRELATION

Many investigations, especially as seen in medicine, are attempts to establish a relationship between two different variables. A *correlation coefficient*, represented by the letter *r*, is used as an index of the linear relationship between two variables. Correlation coefficients show how strong a relationship is, and they range in value between –1.0 to 0 to +1.0. Negative values represent negative correlations, meaning that as one variable goes up, the other goes down. Positive values represent positive correlations, meaning that as one variable goes up, the other also goes up.[5]

DECEPTION AND CORRELATION

What if a person on TV raises his or her hand and tells you to raise your hand every time he or she does so? It might appear that the person on TV is controlling you. What if you were then to find out that the person on TV is actually raising his or her hand when someone outside the window of the studio (whom you cannot see) is doing so? The person outside the room therefore is, in actuality, controlling your hand raising. This example demonstrates the weakness of correlations. The person watching TV and everyone else who sees that program represents the scientific community—that is, that group that observes and assesses the study from their point of view. It would appear to them that the TV personality is causing the hand raising.

Correlation is used very frequently in scientific research and makes exciting claims. However, correlation shows patterns, but *it does not imply causation*. The actual cause in the example above was the variable outside the TV studio and not the TV personality. In science, variables that are not able to be seen by simple observation may be the cause of correlational results.

Because correlations do not explain cause and effect, other variables may intervene. I once heard an outlandish scientific statement: "Women who drive Mercedes cars are less likely to die from breast cancer." Does a Mercedes-Benz actually fight breast cancer? How silly. Obviously, females driving expensive cars probably also have more money and thus greater access to health care and preventative medicine.

In this way, their survival rates are better. The correlation is hyperbole, but it makes the point that correlational research can be deceptive.

Because correlational studies are used extensively in medical research, the powers behind many medical conclusions are weak. It is unethical, in many cases, to conduct experiments on human subjects, so ANOVAs and F-ratios are not suitable tests. Instead, comparisons of groups, without controls, lead to correlational reporting and often irresponsible conclusions.

Simply looking at data from a host of angles allows many correlations to be reported, such as the Mercedes–breast cancer link. Thus, the correlation is a mathematical method by which many of the pseudosciences gain access to credibility. Oddly, I got my first college teaching job (I found out after the interview years later) because I was a Taurus and the other candidate was a Libra. So as logic goes, if a large number of people with the Taurus sign show stubbornness, this becomes their characterization. While entertaining (and I must admit that I am known to look up my sign), correlation is weak and often nonscientific.

"Almonds are good for HDL (good) cholesterol numbers," "Bypasses are better for longer term survival of cardiac patients than angioplasties," and "Atmospheric mercury deposition tends to be higher downwind from large industrial areas."[6] These statements show interesting trends and are exciting, but they are also based on correlational research. This is not to negate the importance of such findings. Correlation is an excellent and effective means of gathering a large amount of information to show trends. However, there are dangers in drawing conclusions from correlations. Unfortunately, such pronouncements are popular and give instant assumptions for the media to fixate upon.

SCALAR TRANSFORMATIONS

Scientists may use other methods to skew the appearance of their results. Scalar transformations are changes made to scales on a graph that bend the appearance of data. Researchers may (consciously or unconsciously) misrepresent the findings of investigations in the presentation of data. This is not considered acceptable in the scientific community, but it happens more often than one would expect. To illustrate, Graph 1 in figure 3.4 shows a weak relationship for the results because the units are larger than in Graph 2. Graph 2 alters the scale of the y-axis, making units smaller, thus exaggerating the slope of the line.

Graph 2 gives the impression that a relationship exists that, in reality does not. It exaggerates the slope to make the public think there is something happening. Climate change research, a hot-button political topic, often shows data (for both sides) with scales that, let's say, favor one side. Showing temperature increases over the past century without showing the many fluctuations over geologic time (prior to human history) gives only a partial picture. This kind of deception casts doubt on a study's credibility.

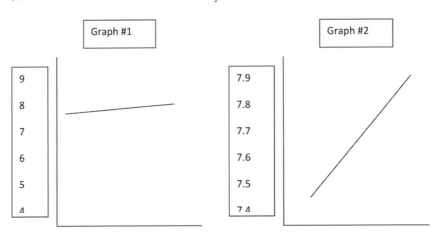

Figure 3.4. Scalar transformations between two graphs.

INTERPOLATION VERSUS EXTRAPOLATION

Another misuse of mathematical representation occurs when data are expanded upon from the graph. This occurs with *interpolation*, which is defined as "filling in" data points in between existing dots in a table or graph. Whenever a line on a graph is drawn, an interpolation is made to fill in data points by making the line. In the example of helping behavior given in table 3.1, a plausible set of data is given based on studies of muggings. When a mugging situation occurs on the street, the number of onlookers affects whether or not people around a victim will actually help that victim. An interpolation based on the data in table 3.1 would suggest that there would be four helpers for every seven onlookers. This is the number of helpers estimated to be in between five and ten onlookers. However, we really do not know, and interpolation is really speculation.

Conversely, *extrapolation* is predicting beyond the data presented in a graph. It is based on trends revealed by the data but makes predictions beyond the limits of the data. Extrapolation is risky because the study has not obtained numbers for the claim. Predicting beyond the graph is guesswork and highly speculative. In economics, performance in the stock market cannot be based on an extrapolated future result—and this is a stock market caveat that has lost many a person his or her shirt!

Extrapolation produces exciting results sometimes and sparks interest in future research. It leads to future investigation to determine if predictions hold true. Consider the helping behavior data set in table 3.1. If we extend the table downward, with even more onlookers, and predict that zero helpers would come to the victim's aid, it has applicability to a common situation (if mugged on a busy New York City sidewalk, don't expect any help). Further studies on helping might likely be stimulated. Extrapolation is, however, untested, and unverified data predictions are a dangerous tool of the pseudoscience.

Table 3.1. Helping Behavior

Number of Onlookers	Actual Helpers in Mugging
5	5
10	3
15	2
20	1

QUALITATIVE RESEARCH

Some studies may be best performed using methods that rely on observations and real-life situations, not numbers. To illustrate, psychologist-researcher David Rosenhan published "On Being Sane in Insane Places" in the journal *Science* in 1973. This study examined the effects of labeling patients within psychiatric hospitals in the late 1960s. The Rosenhan experiment explored the validity of psychiatric diagnoses at various mental health facilities across the country.

Rosenhan used eight individuals with no history or hint of mental illness. Each pretended to have auditory hallucinations to gain admission to twelve different psychiatric hospitals in five different states in the United States. They were essentially "pseudopatients" (three women and five men) who were instructed by Rosenhan to feign hallucinations to gain admittance. Then they were instructed to act normally throughout the study.

Surprisingly, all of the pseudopatients were admitted to the hospitals, and staff at the hospitals determined that each patient had a mental illness. The study showed the dehumanizing effects of being admitted into a mental institution. Staff openly talked about patients in front of them as if they were not there, randomly searched their personal belongings without cause, and labeled many of their normal behaviors as abnormal indicators. In fact, none of the staff suspected that the pseudopatients were imposters because of the strength of the placed mental illness label. Pseudopatients actively and openly wrote copious notes about their study experiences without any staff suspicion. For example, one nurse noticed the note taking by one of the pseudopatients and claimed that her "pathological writing behavior" was problematic.

At the start of Rosenhan's study, pseudopatients were given the goal to get released from the hospital. Even though they were well and acted normally in the hospital after admittance, pseudopatients were held and labeled mentally ill. They were primarily diagnosed with schizophrenia, but some with manic depression. The participants spent many weeks and even months in the hospital. Ultimately, in order to gain release, they were each forced to admit to having a mental illness and agree to take antipsychotic drugs.

One of the hospitals was so offended by the study that the staff challenged Rosenhan to send pseudopatients to them so that they could successfully disprove the study by identifying the imposters. Rosenhan agreed and in the following weeks, out of 195 new patients admitted to the hospital, the staff identified forty-one as impostors and suspected forty-two more. Of special note here: Rosenhan, in fact, sent no one to that hospital. The study thus concluded, "It is clear that we cannot distinguish the sane from the insane in psychiatric hospitals." This exposed the effects of psychological labeling and dehumanization in the mental health professions.[7]

As occurs in all scientific debate, the opposing side presented its counterarguments to the study. The psychiatric community responded with the criticism that the design was based on false claims and deception. If a patient enters a doctor's office complaining of chest pains, it is prudent for the doctor to do a series of cardiac tests. Similarly, psychiatrists work off patient symptoms and if the patient is lying, the diagnosis is naturally affected. Some might even charge that Rosenhan was unethical in his use of deception to debunk mental health science. Controversial but telling, the study could best be done using these techniques. Observation and ethnographic analysis were the right methods for studying the structure and culture of mental institutions. Quantitative methods and statistical analyses would have been inappropriate for studying hidden structural defects in diagnosis within the psychological community. Surveys and data would have been useless in this example.

Qualitative data analyses, as used in Rosenhan's research, have foundational importance in adding richness of detail to many scientific studies. *Qualitative analysis* is defined as the reporting and use of data that are non-numerical. This type of analysis usually involves deeper study of fewer subjects. Therefore, it tends to lack the generalizability of most quantitative studies. However, its strength lies in adding depth in ways that quantitative analysis cannot.

To compare the two types of analysis, consider the earlier bathe-or-not example. A single study may investigate a nonbather and the effects on his or her lifestyle, chronicling skin and psychological health, thus gaining deeper insights into the whole person and the whole phenomenon. It is more of a gestalt approach (considering the whole), with specifics brought together to yield a larger picture. A study of such sorts might look into the patient's personal life, journals and diaries, and effects on the family life from not bathing and not conforming to the norm. Quantitatively analyzed, the individual may be placed into a "success" group by having good skin and being less susceptible to MRSA. Qualitatively, however, the person may have suffered terrible sleep deprivation and depression resulting in a diminished quality of life due to societal rejection for not bathing. Only through a more in-depth and qualitative approach would these kinds of results be reported and considered. This illustrates the strength of the qualitative analysis.

Naturalistic observations are studies involving the natural setting of a scenario. These include keeping a journal, taking copious notes, and interviewing as methods of qualitative data collection to investigate both individuals and whole groups. Sometimes, as in my own research on the attitudes of first-year college students in science, focus groups are conducted in which subjects meet in an organized manner to discuss certain topics. This allows the researcher to observe subject interactions in a seminatural setting. In this way, individuals may be studied and further explored. Data sets from qualitative studies can be processed through a systematic method of coding analysis for detecting patterns and making conclusions.

SCIENTIFIC MODELING

Many unknown phenomena cannot be directly studied. Scientific models, however, can be formed, which are a representations of how something works. Models are a simulation of the real world. Often deriving from an investigation or series of studies, models help in making predictions about how something will work by extrapolating from the artificial condition to the real world. Global climate models and even weather forecasts give predictions of how the atmosphere will behave in the future. These predictions are based on mathematical or artificial situations and are thus nonempirical. This does not mean that they are classified as pseudoscience. Quite the contrary: they are based upon assumptions grounded in scientific research and are the best alternative to empirical science. Many astronomy and physics areas deal only with models because they cannot gather evidence from the extrapolated hypotheses. Einstein proved relativity and nuclear fission on paper (in 1905) using mathematics long before the atom was actually split in the 1940s.

Modeling is an important part of science but is limited due to its detachment from the real world. For example, cancer cures in mice, with antiangiogenic drugs are abundant. In humans, however, cures are less successful. Using mice as a model for human reaction doesn't always hold true in the real-world application. Models developed from research on animals and in the lab are weak because they are often not accurate depictions of how human bodies will react. Despite this, models are necessary for guiding research. The quote below by Kenneth Paigen expresses this.

> People have a basic misunderstanding of the mouse. They get upset that the exact pathophysiology might be different, but a mouse is not a total mimic. A mouse is a discovery tool, a device, to understand the molecular pathways that underlie disease processes. You can't do a lot of types of experiments on humans. You can't order people to mate . . . We have to turn to models.[8]

NONEXPERIMENTAL RESEARCH

This chapter began with a plausible experimental design to study neti pots. However, many studies are nonexperimental. By that I mean that they do not get set up by isolating an independent variable and having a control group. Instead, they are descriptive, answering questions about natural phenomenon. Nonexperimental studies are not meant to test predictions or refute hypotheses. Instead, they might explore the coast of Maine for levels of mercury, describe moon soil and topography types, or count species in the Antarctic peninsula. They gather information and are valid and fundamental science. Descriptive research furthers our understanding of the world and universe. Nonexperimental research is foundational in the history of how science developed, as will be discussed in the next chapter.

4

Science for Every Person

The whole of science is nothing more than a refinement of everyday thinking.

—Albert Einstein

AAAS/NSES STANDARDS

In recent decades, the movement to help all people learn and be able to do science has been a national focus. Science is not something only done by scientists; it is needed in everyday life and uses the common sense people already have. One thinks scientifically every day. The previous chapter points to ways of refining that thinking. A goal of any civilized society is to bring science to every person to make his or her life better.

Scientifically literate people not only have better quality-of-life decision making, they also contribute more to the economy. The ability to calculate square footage and figure out the best type of flooring for a house is a skill that involves science, for example. A good carpenter uses science and scientific thinking to consider many variables, in order to do a better job. The wrong flooring, excess material ordered, and poor engineering layout all cost the customer money and lead to waste. A scientifically literate population makes for a stronger nation. Thus, a goal of the U.S. government's Department of Education has been to formalize and assess what should be learned in the form of science standards.

Guidelines for science learning have been developed both nationally and at state levels. Putative agreement is shown in a couple of documents: Project 2061's Benchmarks for Science Literacy, developed by the American Association for the Advancement of Science (AAAS), and the National Research Council's National Science Education Standards (NSES) which are both highly consistent. For the most part, the National Research Council relied heavily on the benchmarks in drafting its content standards, as stated in the introduction to the NSES documents. NSTA (National Science Teachers Association) extended the documents to include standards for science in higher education.

These documents are guides for the development of state framework committees, school district curriculum committees, and developers of instructional and assessment materials. The focus of these national standards emphasizes the importance of major themes and ideas in science over memorization of vocabulary. An information explosion (via the Internet as well as a proliferation of journals and data) has led to a glut of knowledge. This requires an organization of scientific thinking into a set of foundational areas to focus all of that information.

Benchmarks and NSES standards both represent very involved years of work by experts in science and education. The work has produced a valuable consensus in guiding a program for bringing science to all Americans. It is not merely the science elites that the standards seek to educate. Mathematics, science, and technology have come together to bring about products of scientific inquiry, requiring a scientific literacy that has become necessary for everyone. Everyone needs a base of knowledge and skills (both intellectual and practical) to use science products, debate current issues, and make informed decisions.[1] Should a person merely accept a plumber's word about his or her piping or be able to understand pressure, friction, and materials to judge the quality of the plumbing job? Science literacy empowers the individual in a plumbing repair case, but it also seeks to educate about larger health issues and informed decision making in communities.

Scientific literacy is also growing in importance in the workplace. The job world demands less rote memorization of facts and places a greater emphasis on problem solving, being creative, designing plans, and forming solutions to emergent problems. The National Science Education Standards presents their vision of a scientifically literate populace. Major themes of the National Research Council's vision and content emphases will be outlined in this chapter. They discuss what students need to know, understand, and be able to do to be scientifically literate at different grade levels. The intent of the standards is to create "science standards for all students."[2]

While the goal is admirable, it is not without contestation. Many also feel that either some students are unable to learn science or that some should not need to learn scientific ways of thinking. It is true that science is a difficult field to work in, and students cannot achieve in science without the support of "skilled professional teachers, adequate classroom time, a rich array of learning materials, accommodating work spaces, and the resources of the communities surrounding their schools. Responsibility for providing this support falls on all those involved with the science education system."[3]

SCIENCE AS PROCESS

A student may be told a number of factoids, but actual understanding usually emerges through active involvement in the process of science. Implementing the standards will require major changes in much of this country's science education. It will require a refocus from recall and declarative knowledge to inquiry and the scientific process. "Learning science is something that students do; not something that is done to them. 'Hands-on' activities, while essential, are not enough. Students must have 'minds-on' experiences as well," according to the National Science Education Standards.[4]

Science is a process and requires the development of certain abilities: observing, inferring, and experimenting. The implementation of "inquiry" into this process is crucial to scientific thinking and learning. *Inquiry* is a vague term but generally involves inductive inferences made from observations (for example, students describe objects and events, ask questions, construct explanations) as well as testing those explanations against current scientific knowledge.[5] As described in the Typology of Argumentation section in chapter 2, higher levels of thought and reasoning constitute effective inquiry. These levels include use of reasoning skills, discussion and dissemination of ideas, and consideration of counterarguments.

Current standards call for a teacher to not simply focus on one strategy or another to develop science literacy. Sometimes lectures and proscriptive procedures are appropriate. At other times, open-ended inquiry engagement and discussion work are beneficial. The standards ask for a multipronged approach to developing science thinking. Regardless, "knowing" and "doing" in science are based on certain categories. The NSES content standards include the following unifying concepts and processes in science: science as inquiry, physical science, life science, earth and space science, science and technology, science in personal and social perspective, and the history and nature of science.[6]

UNIFYING CONCEPTS

As mentioned, the explosion of information does not allow for a "knowing" of everything in science. This book investigates science through debunking myths, pseudoscience, and cases and examples. It does not merely "cover" all the standards. Science teaching should mirror such an approach to help students engage with the material. That content should be organized around NSES themes.

NSES shows that there are unifying concepts and processes within each of the content standards to include 1) systems, order, and organization; 2) evidence, models, and explanation; 3) change, constancy, and measurement; 4) evolution and equilibrium; and 5) form and function. There is a certain order and organization to the universe, studied as far back as Aristotle. However, science is about change and measuring those changes, with examples including evolution of both living creatures (for example, humans) and physical systems (for example, solar system formation). All objects and energy take some form and perform some function. Thus, the final concept unifies with the others to show the innate link of the ideas. In the next section of the book, science topics will be explored to discuss these themes in everyday life.

Obviously the NSES standards are geared toward teaching students. However, all people can view science thinking in terms of these major themes. While the integrative processes described should bring together students' experiences in science education classrooms across grades K–12, they are truly useful ways of thinking about almost every concept in real life. Using plumbing as an example, dealing with a leaky faucet may be viewed as disequilibrium of a system. It requires the development of a model for repair, a visualization of the problem, and an inquiry to ask the right questions for devising and implementing the solution.

The unifying concepts and processes of the NSES standards are the central point of this chapter: to organize science knowing in a way that makes sense to everyone. Studying interesting and applied aspects of science is the best way to learn science. In chapter 7, "Pseudoscience," and chapter 8, "Debunking Science Myths: Separating Fact from Fluff," science is presented using this approach.

CULTURAL TRANSITIONS

Do you recall an experience from your own science classroom days? Is it good or bad? Do you have a child or friend entering college as a first-year student, considering science as a major? The world of education in science is different from the real world and very different from learning in other fields of study. This section is vital for parents and students interested in starting a career in science. It describes and makes recommendations about transitioning to college science in particular. It is at this transition that more than half of people interested in science precipitously drop out.

Science has its own way of thinking, which can be applied to any circumstance in a real-world setting. However, think about when people become acquainted with the scientific way of thinking. It is most often during their school experiences that students become incorporated into a culture of learning qualitatively different from the real world.

First, science experiences in school are often done within an artificial classroom experience, with procedures to follow in a laboratory and tests to show that something was learned. It is often not in situ, so students feel detached from the relevance of the topic learned. Sometimes such structure in teaching is important and necessary, but often, traditional science classrooms are detached imitations of real life. A nonengaging lab may ask for simple rote memory of bones and muscles. Alternatively, bones and muscles can be viewed on models, on peers, on patients, and by using technology or dissection. Laboratory experiences do not need to be devoid of the real world, but they often are. This is the unfortunate expectation in traditional science learning.

Interest in science is documented to decline substantially as a student's time in school progresses. This phenomenon is termed the *leaky pipeline*. It is during the transition from high school to college science that science interest really takes a hit, with as many as 50 percent of students leaving science areas during their first year of college.

In general, a higher education institution's ability to retain students is affected in part by the retention of students within their majors. This is why colleges are so concerned about the retention of students even in difficult areas such as science and mathematics. While the retention of students in a particular major, such as biology, does not necessarily directly equate to retention in the institution, the chances of holding on to students in the organization increases when students remain in their particular chosen major.[7] Thus, for example, if students remain biology majors after completing the introductory course, there are greater chances of retaining those students within the college.

Retention expert Vincent Tinto proposed three ways of viewing first-year college student retention, which can be applied to the sciences: the psychological, the social, and the organizational perspectives. The first two models involve a personal and group adjustment, which are a part of the organizational view. Everything the student experiences takes place within the organization. Thus, this perspective will be discussed as a way of understanding how students adjust to college and college science.

Not only the sciences but also college as a whole are an adjustment challenges for people. As a result, both science and college may be viewed as a challenging cultural transition. Of Tinto's three models, only the organizational (institutional level) paradigm allows for the development of a research program that furthers both individual and institutional empowerment in an attempt to increase science student retention. Through the exploration of this perspective, solutions can be developed to improve both an entering student's chances for success and an organization's retention in science. Tinto points out that "if there is a secret to successful retention, it lies in the willingness of institutions to involve themselves in the social and intellectual development of their students."[8] It is the conceptual framework offered by this viewpoint that is most optimal for understanding the science student experience.

THE ROLE OF FACULTY ASSUMPTIONS WITHIN THE ORGANIZATIONAL PARADIGM

First, what science faculty expect of students entering into their classrooms is the most important facet in retaining students' interest and thus, their academic success. A proper match between high school and college preparation is essential for success. Second, the school should make the overall transition to a college environment better for students. Third, science, as a unique way of learning, should be recognized and addressed by the institution. The following section looks at the impact the educational institution has on satisfaction, socialization, and therefore, the success of science students at a college or university.

The central tenet has been that departure is mainly a reflection of institutional behavior rather than of individuals within the institution. Researchers have looked at the effect of organizational dimensions such as size, faculty-student ratios, resources, goals, and faculty instruction on student success. Student success in science is thus partially dependent on certain organizational dimensions.

Appropriateness of faculty requirements to match the knowledge, abilities, and dispositions of their incoming students with their preparation is often beyond the control of the organization. Those requirements must be appropriate given the academic preparation of students during their secondary school experience. However, faculty expectations are often unstated and unknown. Their expectations should be studied to improve the match between high school and college science.

Also, too often there is pressure to simply "make things easier" and pass more students, without improving the congruence of preparation and expectation. Science courses are notorious for failing large numbers of students. There are many reasons for this, but when a student does not meet expectations, the goal of the institution is to help them to do so—not to lower standards.

ORGANIZATIONS

The organizational model should be of interest to institutional planners concerned with the improvement of retention within their schools. The variables studied in this approach can be administratively altered. It is possible, for example, to develop programs to smooth the transition between high school and college science programs. Entrance exams, remediative courses, and communication with secondary schools might be vital in improving students' science learning.

Analyses of institutional retention are being done fervently in higher education across the nation due to the extension of NCLB (No Child Left Behind) to higher education. A focus on the sciences has also been embraced through many new STEM (science, technology, engineering, and math) grants. However, very little research is being produced to define and improve the congruence of faculty expectations between secondary and postsecondary science programs.[9] My research findings will be presented in chapter 11, "Getting People to Love Science," which will offer solutions to the science retention problem.

Immediate academic demands that students face in the science classroom are most important in retaining students in science. Faculty *assumes* students should know and be able to do certain things to succeed. My research has shown that those assumptions are quite different from students' high school science preparation. Organizations need to further investigate such transitions to better integrate students into the academic college culture.

Culture matters: the importance of both academic and social integration for first-year science students into the institution's culture has been shown to contribute significantly to student academic success during the first year of college (see Tinto, 1993; Wolfe, 1991). Students are very concerned with their social lives and need to find a place in college that fits their personality and goals. If smelling like amines after organic chemistry lab makes a student feel like an outcast in the cafeteria (this author has felt it), then these concerns should be addressed. Amines do smell awful, and I lost a few nonscience friends that way (joke).

Students' adaptation to new cultural norms and values is a part of their success during the first year of college. Because of this, the organizational model of retention should be understood in terms of a sociological and thus a cultural change. A diagrammatic representation of this role of faculty requirements in retaining students within a college is included in figure 4.1. The following section elaborates upon this mechanism illustrated in the figure and should be used to accompany its reading.

TRANSITION AS A RITE OF PASSAGE

The process involving a student move from one level to another may be viewed as a cultural transition. It should be understood through drawing from the field of social anthropology, which studies the rites of passage in tribal communities[10] and immigrant experiences to new cultures.[11] This section outlines a conceptual framework for viewing that transition from organization to organization.

A social anthropologist, Arnold van Gennep (1873–1957), proposed a model describing the rites of passage to membership in tribal societies, which easily compare with the first-year college student experience in science. Van Gennep was concerned with the movement of individuals and societies. His research examined the movement of individuals from membership in one group to another, similar to the ascent of students from high school to college communities. The problem of becoming a new member of a community is conceptually similar to that of becoming a college student. The process of institutional persistence in science, therefore, involves three major transitions, which are in line with the model. These transitions are called *rites of passage* and are marked by three phases: separation, transition, and incorporation, each having special ceremonies and rituals.

In cultures and in college, people experience a change in their way of thinking. There is an orderly transition of beliefs and norms from older communities to the replacements. The process allows new members of a community to become assimilated into their new society. This model of transition is applicable in explaining the kinds of changes students undergo during their movement into college life.

Vincent Tinto points out that "however close, the life of families and high schools and the demands they impose upon their members are by necessity qualitatively different from those that characterize most colleges."[12] Thus, in order for a successful transition to college to occur, associations of past groups must undergo a period of separation. There is often a decline in interactions with former groups and a rejection of the views and norms that characterize those groups. Just as immigrants must separate from former societies, the students' retention in college science depends on their leaving their former communities.[13]

The second phase, the transition period, follows. During this time the person interacts with new members of the community to begin learning the new views and norms characterizing the group. In science, the new group has a scientific way of

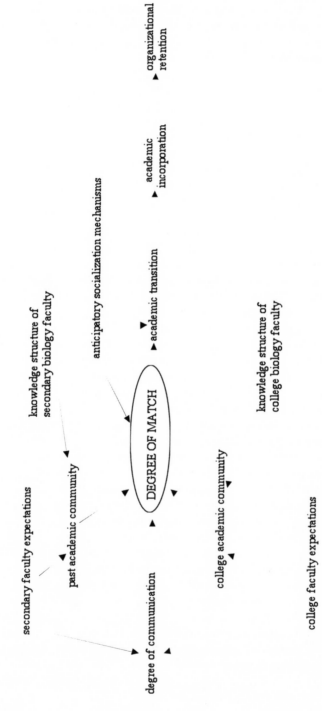

Figure 4.1. Framework for successful academic transition for majors in introductory college biology.

thinking associated with it. To ensure the adoption of new ways and a separation from old patterns, mechanisms such as isolation from past group members and training to learn new values are used. It is at this point that college science programs should intervene and enculturate new students into their community.

When considering this transition period in the first year of college science, successful students replace high school values and norms with the academic, social, and psychological knowledge, abilities, and dispositions required to succeed at the college level. By its very nature, the center of the academic and social communities is the classroom. Faculty and students interact in the introductory science classroom, and for the first-year student, it is the first real exposure to the norms of the new community. It is at this level that the process of transition is so important. The broader process of incorporation emerges from the students' classroom-based involvement with both peers and faculty.[14] Some studies show that it is during the first weeks of college that forming social connections in the introductory classroom serves as a precondition for students coping with the academic transition.[15]

The third phase of incorporation involves the successful practice of new patterns of interaction in the new group and the ultimate establishment of a competent membership in that group. A fully integrated member can once again begin to re-establish past group associations, but only as a member of the new group. College students who complete the incorporation phase gain strength, power, and confidence as time progresses within the new community, as shown by a variety of educational researchers. In science, incorporation is vitally important, with students coming to identify with their subculture. Science people become their own community, as will be discussed in chapter 9, "What Are Scientists' Responsibilities?"

A major obstacle to a successful transition is isolation. Being cut off from former communities and alone in a new setting, college science students rely heavily upon their classroom for acceptance. This is why high rates of failure in introductory science are so devastating for overall student success in college.

For students, the movement into college science requires change from a high school position as a known member of one group to that of a stranger in a "new college community," a state that Tinto terms *temporary normlessness*. It is at this time that students are most likely to drop out of the college science community. According to Tinto's research, this is because students feel most uncomfortable with the loss of their high school value systems and yet are still without the support of the college community.

One of the major transitions from high school to college involves the "unlearning of past attitudes, values, and behaviors and the learning of new ones. . . . In their place new identities and interpersonal networks must be constructed, and new academic and social structures, attitudes, values, and behaviors must be learned."[16] Thus, in general, when using migration as a metaphor, "human development itself can be seen as a succession of migrations whereby one gradually moves farther and farther from his first objects."[17]

INTERVENING VARIABLES

Many different variables interact to make the dynamic of college science student success more complex. For example, some students may be successful without going through these phases or may experience only partial stages and/or phases in different sequences. The stresses of separation for first-generation college students were studied, showing that the feelings of isolation and anxiety for this group were more pronounced than those in non-first-generation college students.[18]

Commuters do not need to separate entirely from former communities. While students attending nonresidential colleges feel an initial ease during the separation phase, the task of persistence in new science communities may be more difficult for them in the long run since "a person's ability to leave one setting, whether physical, social, or intellectual, may be a necessary condition for subsequent persistence in another setting."[19]

In other situations, the experience of transition depends on the social and philosophical character of one's past communities. A study of first-generation college students shows that a lack of emotional support for higher education from family or past peer groups makes the transition to college science significantly more difficult. The separation phase often requires a visible rejection of the values held by family and peers for successful incorporation into the college community. However, the importance of positive parental encouragement and support, as seen for individuals with college-educated families, increases student retention and eases the transition to college. The disadvantage of first-generation college students is also demonstrated by various studies of such students describing their expectations for college success as different from non-first-generation students. The first-generationers held a more dogmatic view of college and expected to be told, didactically, what they should do to be successful. This is more characteristic of the lower socioeconomic status (SES) parental guidance style to which the students were likely exposed: authoritarian. Successful resocialization to college for these students required a change of these dogmatic academic expectations to a more flexible approach to learning.

Similarly, research on foreign students, rural disadvantaged, and students ethnically or religiously distinct from the majority show similar difficulties during the transition to college. In general, individuals coming from families, communities, and schools with norms and behaviors different from those of the college science community face greater challenges in achieving successful incorporation into the college science structure.

The ability to successfully integrate into college life is determined, in part, by the degree to which individuals have prepared for the process of transition prior to formal entry into college. This is a main purpose of this part of the chapter: to prepare incoming students for college science.

In considering the transition between high school and college, there remain questions as to what kind of preparations actually help students for a successful incorporation into college science programs. Studies show that prior understanding of college life and science education is helpful.

In using van Gennep's model of the rites of passage to understand the first-year college science student experience, a sensitivity to the differences in characteristics of the college science community with those societies studied by van Gennep should be considered. To illustrate, student membership in college communities is almost always temporary, with the goal of leaving successfully always held in high regard. In contrast, van Gennep's communities in emerging third-world nations were more permanent, with social structures persisting for possibly many generations. Also, van Gennep's incorporation phase emphasized the importance of formal rituals and ceremonies as demarcations between phases, whereas in application to college transitions, this is for the most part absent (aside from formal rites of passage in Greek life and orientation programs). While the van Gennep model does allow for a general way of thinking about the informal processes of college science student interactions that lead to incorporation, successful student retention is more complex than a single process of resocialization.

TRANSITION AS AN IMMIGRANT

The first-year experience and the immigrant adjustment are, generally, congruent to each other. The student must transition, just as an immigrant, into a new culture of learning in college. This includes culture shock to new ways of studying; new language acquisition (big words such as *glycolysis*); new academic and social requirements; and the internalization of academic/cognitive, bureaucratic, and social norms. The academic community has recognized the transition of students from high school to college as problematic.

The first-year college experience is, on the whole, a strange migration. It should be understood by those teaching and developing policy for these students. A parallel between immigrants and new college science students may help shed light on some shared challenges with immigrants.[20]

SCIENCE STUDENTS AS IMMIGRANTS

The immigrant analogy "is apropos precisely because it contains so many of the same elements encountered by students when making the transition to college."[21] For example, the immigrant and student populations have different dimensional characteristics when compared with the populations as a whole. Generally, both are younger and more adventurous, more flexible, and open to the change needed for resocialization. Both face a dominant group that rejects new members. Further, both hold goals for improving their social and/or economic standing by choosing to make such a transition.

To further the metaphor, the academic and social positions held by both first-year science students and immigrants are relatively inferior to those held by faculty, staff, and upperclassmen and native community members, respectively. Neither has proven to be competent in the eyes of the dominant group. They compose, in a sense, a minority group position.[22] "Like immigrants who arrive with a lack of understanding about the structure and processes of government," first-year students completely lack knowledge of the governance structure or power system operating at the college level.[23]

Consider the possible scenarios for the first-year student's initial experiences in the new "country." Upon arrival, a myriad of academic and social tasks are encountered that could be difficult to manage (especially if nothing has prepared them in his or her "native land," which is the high school in this case). At first, the student's new freedom from parental and high school control appears fantastic, and the tasks are put off. The research shows, however, that academic culture shock sets in generally in the mid-semester after midterm grades are reported. Both academic and social aspects of students' self-concepts decline during the first year. A host of plausible explanations are available for each individual, but the research shows general declines in academic trends as the first-year student progresses into the year. Perhaps feeling pangs of separation or an anomic sensation with the absence of previous social networks is at work.[24]

The rules to succeed for both the immigrant and student change drastically in the transition. First-year students enter college expecting merely a more sophisticated academic version of high school. The academic culture in science in particular is both shocking and frustrating to the new "immigrants." Student access to professors is much diminished compared with secondary teachers. New students perceive faculty as aloof (for example, citing that faculty do not often learn their names and do not keep regular office hours).[25]

A variety of other academic differences in the undergraduate science structure have been shown to frustrate incoming students. Although rarely encountered in a college classroom, quizzes, extra-credit assignments, and unit exams are common high school faculty methods of assessment expected by first-year students from their "homeland." Ironically, many college courses offer only two chances for assessment, usually a midterm and final examination or a research paper.

The changing assessment is a single example of a larger change in the academic culture for the immigrant student leaving a more nurturing homeland of high school. The close supervision of secondary students (for example, homeroom, attendance, late slips, hall passes, class period bells, and study hall) is replaced with a lack of daily structure and a demand for academic self-discipline.[26]

In concurrence with the changing academic rules, just as the immigrant often must learn a new language in the new homeland, the science student is faced with the new language, "academese," of higher education and science jargon. In the science classroom, "some professors appear to be word imperialists who continually oppress students with a vocabulary that is totally incomprehensible to the new immigrant."[27] Just as the dominant culture expects the new immigrant to learn the new language, so does the college professor. *Mitochondria, empirical,* and *inductive* are words within these chapters, but they are often a fright for those entering science students. On the administrative side, the bureaucratic maze of rules, regulations, and paperwork are difficult to navigate. Terms such as the *registrar, bursar, FAF, EPY 240*, and *CLEP* present obstacles to first-year student functioning.

The changes in learning demands for college sciences described in the immigrant analogy have been shown to cause a variety of problems for the science student as immigrant. Some reports show that college students cannot meet the increased levels of sophistication not found in high school science; namely, the requirement for students to present and support complex arguments. College science seems to demand a move beyond the literal mode of high school thinking to more abstract thinking. Logistically, as well, first-year students view new college science expectations as ambiguous: the syllabus is often not followed, daily assignments are not given, and assessments are usually conceptual rather than fact-based.[28]

In one study, a college science instructor complained that high school had taught students "intellectual bulimia—stuffing themselves with volumes of factual material and then regurgitating it back for the professor."[29] The changed emphasis on improved reasoning and integrative thinking is a common transitional theme in the move from secondary to postsecondary science, as described in chapter 2, "Science Is Arguing." This chapter should help students to argue science better through understanding the philosophy and history behind this sort of thinking. Knowing the challenges college science students face should help better prepare them to succeed.

THE MECHANISM OF REACHING NEW LEARNING DEMANDS

The most challenging academic feat for students is learning what the professor expects of them to be successful in his or her course. Certain assumptions about reading and writing ability, study skills, academic-management methods, and even dispositions are clearly different from the requirements of the secondary teacher. First-year students generally come to class expecting that the professor will "tell them what they need to know," as did their high school teachers. This student-faculty difference in assumptions about what is required for success can be a major block to improving student retention.[30]

Although the process of incorporation into the college environment parallels the acculturation that immigrants experience in a foreign society, unlike the immigrant, the first-year science student expects the new environment to be similar to the one from which the student has immigrated. It is this fallacious set of assumptions that greatly hampers the student's transition.

SCIENCE STUDENT EXPECTATIONS

Requirements perceived by both immigrants and first-year college students are often mythical. For example, just as immigrants expect that work will lead to wealth, students also "bring with them a set of myths about what they will encounter in their new homeland." Both immigrant and student are often very disappointed. The most dangerous of those myths is the one affirming that current secondary school preparation will prepare them for college life.[31]

These preparations are, in part, a result of what high school teachers think college faculty will expect of their students. Requirements that secondary school teachers have for students comprises knowledge, skills, and dispositions to engage in college science. The research shows, however, that high school teachers may not know what college instructors expect of the students entering undergraduate science.[32]

The requirements of college faculty are defined as the prerequisite knowledge. Prerequisite knowledge is the set of skills and dispositions that students must possess to be able to successfully engage in college science learning activities. The lack of publicly defined postsecondary faculty requirements in the sciences may contribute to the academic transition problem for students in the first year of college science. Accrediting agencies for colleges are now looking more closely at assessments of learning. This is an important shift from allowing college science faculty full freedom to determine goals, standards, and objectives. The extent to which college faculty should be circumscribed by state, national, or institutional standards is debatable, and academic freedom is important. However, when the high school teachers preparing students for college science do not have established standards at the college level to act as a guide in their instruction, student preparation suffers.

COLLEGE SCIENCE STANDARDS

Developing acceptable requirements in college courses is a first step in preparing secondary students so that their knowledge, abilities, and dispositions entering introductory college science will match the academic demands of the undergraduate faculty. The Benchmarks for Science Literacy (1993) and the National Science Education Standards (1996) nationally define the content, instructional, and assessment strategies appropriate for the K–12 science curriculum. No such set of guidelines, however, has been effectively implemented or defined across the undergraduate science teaching field. No Child Left Behind is a K–16 mandate, meaning that college reform is also required under the law.

While the National Science Teachers Association (NSTA) apply the NSES standards, meant for K–12, to college science, few higher education science faculty are aware of its existence. In the College Pathways series presented by NSTA, all six science standards—teaching, professional development, assessment, science content, science education programs, and science education systems—are applied to postsecondary education. The document is a useful rresource for administration and science faculty guiding their programs and courses to become more inquiry-based and student-centered. College Pathways uses vignettes and cases to show models of teaching for college science courses.

While science standards are documented in numerous places (for example, California Higher Education Institute, 1984; College Expectations, 1992; Oklahoma State Regents for Higher Education, 1992; State University of New York, 2000 and institutionally from course catalogues and syllabi), there is vast inconsistency in teaching and assessing science courses at the college level.[33] Thus, the definition of what college science entails is still elusive.

One professor can teach chemistry with a completely different focus as compared to another. Also, the difference between high school and college science courses is vague. Further work is needed to clarify these distinctions and avoid unnecessary duplication of efforts and unevenness in teaching. For example, many times I will teach students who have reviewed Ernest Rutherford's model of the atom over and over in their courses. Yet, because curriculum is so uneven, some students have never heard of the concept. Efforts are repeated for some and glossed over for others. This small example of atomic structure is only one of the many gaps between levels of instruction.

At the state level, in an attempt to define the learning requirements for the sciences, the Academic Senates of the California Community Colleges, the California State University, and the University of California published an often-cited state-level document on science learning expectations: "Statement on the Preparation in Natural Science Expected of Entering College Freshmen."[34] Its emphasis reflects similar requirements expressed by other state and institutional documents. But the work was not followed up with national accepted standards for college science, nor is there an effective attempt to nationally align curriculum with high school preparation.

At an institutional level, the most general assumption for student learning is the college or university mission statement. Although the clear definition of faculty demands for incoming students has been a goal for most institutions, those requirements are rarely defined beyond the local level. Because there is a host of changed academic demands in science at higher education institutions as compared with those of the high school, there should be proper communication of those demands to students. This would involve science faculty specifying the specific outcomes of knowledge, dispositions, ideas, and processes that they expect of students at the college level to high school teachers.[35] When and how is this possible? Through national or at least state-level documents to define these for high school teachers. No recent administration has pushed for these efforts effectively. The hope behind NCLB was to redefine standards and assessments and to align these with college curriculum.

In the absence of such progress, parents and students should prepare themselves for a cultural change to college science that may be rocky. Through preparing mentally for some poor outcomes, having a back-up plan to get tutoring, or seeking precollege preparation in science, the transition could be improved greatly. Ultimately, awareness is the first step in preventing problems during that first, difficult year of college science.

SCIENCE IS BUILT ON MANY PILLARS

Defining science standards across campuses is especially difficult in science because it is built upon understanding a host of other disciplines (chemistry, physics, and biology). It is unlike mathematics, for example, which is sequential, with the depth of content following a more clearly defined and measurable course. In the sciences, the many interrelationships between its branches contribute to a nonsequential nature for many of the topics.[36]

As a result, there is, on the whole, an unclear definition of college-level requirements at national, state, institutional, and even faculty levels. One can take organic chemistry in one school where it is easy or at another school where it is next to impossible. Even within the same college, different science courses are dependent upon the instructor. This lack of uniformity is concerning. I once had a colleague who partied with the students, showed videos in most classes, and held eight-minute lectures. Students were elated for a while, but learning was limited.

This lack of effectively implemented and defined standards in college science has been cited as a possible depressor of student success. While it feels good to have the easy teacher, when the next class comes, it is time to pay the piper. I have had the experience of getting students who, after taking easy versions of classes, enter my classroom only to fail. Expectations to not study and have a good time led to their attrition. Some are able to change their habits, but it is very difficult. Science courses build upon each other and upon other areas. When the pillars are weakened, the whole edifice often comes crashing down. Easy teachers are not the easy answer to improving standards and grades, and parents should resist the temptation to call for such.

Although most institutions and states have written requirements for various undergraduate content areas, the actual implementation of these guidelines is another matter. There is not much oversight on what college teachers do in the classroom. High school teachers focus on their teaching, but the college atmosphere has demands of their faculty for research and service. Thus, the decreased emphasis on pedagogy has effects on making sure that learning demands are appropriate. The question should be "What is the preparation in the secondary curriculum to bring students to a level of integrative, critical thinking that would enable a smooth transition to meet the postsecondary demands for these abilities?" This would help to ease the transition best. Chapter 11, "Getting People to Love Science," will discuss the methods by which students can be better recruited and retained into the sciences.

RECOMMENDATIONS FOR SCIENCE CONTENT

If a national expectation for students is that they become more scientifically literate, then the content must be directed to such examples. A major content concern, which is termed the *constant-volume* problem, emphasizes that there is a fixed amount of time for which to choose an ever-increasing amount of content. Most teachers find it difficult to "cover" all of the topics, especially since the development of so many new areas in science (for example, gene technology, medical breakthroughs).

Thus, a better way to teach high school science is to avoid the repetition of knowledge. This is one important way to solve the constant volume problem. As discussed earlier, there should be a better interface between high school and college instructors to avoid a duplication of efforts. The "vertical nature" of science (that there are prerequisites of understanding in other domains to attain higher-level thinking in any one science) is a main barrier to content determination. Since science curricula tend to be vertically structured, a certain level of content knowledge for incoming college students is necessary for success in undergraduate science programs. Parents and students should know that there is an enormous amount of information that college science faculty think they need to cover. It is impossible to cover everything, and each teacher selects what he or she thinks is most important. This is what leads to the inconsistencies in science course requirements from school to school and teacher to teacher. Unlike those in high school, in which the curriculum is prescriptive, college teachers are rather free.

With the explosion of information in science it is no longer possible to deliver all of the information necessary to prepare students for higher levels. Thus, high school students should be shown how to organize that massive knowledge and search through the content to make educated decisions about science. All citizens need to be able to keep on learning after traditional education ends. There is always new information in science, and by using the tenets of this book, readers should be able to study science on their own. Gaining independent learning styles is valuable for science students throughout their academic careers.

RECOMMENDATIONS FOR SCIENCE LEARNING

Teachers, instead of facing the current problem of what to cover, should help students to more independently take control of their learning. At home and in high school, instruction should foster conceptual understanding so that students can place the information into meaningful networks.[37] Students need to use their real-life experiences to build mental frameworks for making sense of science concepts. Thus, when studying college science, old information affects how students accommodate new information into their mental schemas.

A method of instruction for helping students organize this content is found in explanatory writing. Explanatory writing requires that students organize their thoughts to explain a science concept to a nonscientist friend. Writing can be a powerful tool for expanding, modifying, and developing new knowledge networks.[38] Through writing, they can link together old and new ideas about science. These mental structures would help students better link to the new information and expectations of their college instructors. It follows that organizations should focus on student writing to improve college science success.

Writing helps to emphasize not just what content is taught but how that content is produced and conceptually placed. Students should explore fewer science topics, but explore each in greater depth. Often scientific ideas are complex, demanding that students surrender some prior knowledge from high school and organize their minds around new concepts. Heart structure in high school models leads students to believe that the atria are almost as large as ventricles. Of course, this is not the case, and the atria are very small. It is the cheap high school cartoonlike model that misleads students. Upon seeing dissected hearts, students are shocked. This example should point out that misconceptions can be untaught very easily, but instructors need to be aware of them.

INQUIRY, INTEGRATION, AND STUDENT SUCCESS

John Moore, back in 1984, described science in general as a "way of knowing," a disciplined inquiry into the creation of new knowledge. This inquiry—the way scientists create knowledge, present it to peers, and revise with feedback—acts as a model for students to help them understand science. Instruction in the sciences is most effective "when it captures the methods of thinking that scientists use when exploring the world."[39] Thus, "teachers [should] serve less as deliverers of information and more as coaches to students developing the habit of learning on their own" in the ways scientists do. While the teachers remain the experts, their role is more strategically used to guide the students' own investigation processes, according to John Dewey (1859–1952), an educational researcher in the early part of the twentieth century.[40]

The mere presentation of content in introductory college science courses, without giving students opportunities to see how that knowledge fits within the whole curriculum, does not further the kind of collateral, integrative learning advocated by the national, state, and institutional standards described earlier. Although scientific factual knowledge is needed in the vertically structured sciences, it is only insofar as it can be used in higher-order cognitive processes. Better science thinking compares and connects phenomena of a variety of academic areas. The importance of Dewey's principle of continuity is thus recognized by these recommendations for content in college sciences, whereby students engage in collateral, synthetic learning. This fosters "the most important attitude that can be formed[,] . . . that of a desire to go on learning."[41]

Content learning demands for undergraduate sciences should thus be considered intrinsically multidisciplinary. Rather than focusing on what amount of content is necessary, the organization of that content around themes, issues, or projects can enrich the students' views that the sciences are not separable from other areas of life and can be reasoned about in a more holistic way.

Good alignment between faculty and student expectations for successful engagement in science is based on the establishment of national standards and a standard-

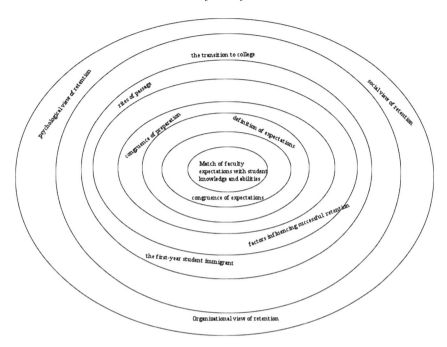

Figure 4.2. The cultural transition to science.

ization of the learning demands for science students. Despite a host of criticisms of standards-based reform efforts effected by the No Child Left Behind Act of 2001, a newer set of reforms has not yet been implemented to guide science education. The research shows that congruence between faculty expectations at the high school and college levels of science teaching is vitally important to the cultural transition students make into college science. This transition is a necessary prerequisite to entrance to the scientific community and should be securely guarded by any administration focused on science education reform. A visual representation of the conceptual organization of the cultural transition to college science and the science community is given in figure 4.2 above.

5

The Role of Critical Thinking

What is dark matter? Is the universe expanding or contracting? Is dark matter causing the universe to expand more quickly than would otherwise be predicted? Is there a good solution to cavity prevention that is better than fluoride? Is a mutant bacteria, *Streptococcus mutans* (with an alcohol gene inserted), able to fight cavity-making bacteria in people's mouths? Are strings of energy making up atoms shown by experimental physics or only shown mathematically? What about quarks?

These are provocative questions currently under study in science laboratories throughout the world. The way in which these questions are asked and answered requires a certain kind of thinking. It is creative and critical of the existing knowledge available to the science community. The way to answer these and other questions is through using critical-thinking skills.

Critical thinking has many meanings to different people but, simply put, it is "the careful, deliberate determination of whether we should accept, reject, or suspend judgment about a claim—and with the degree of confidence with which we accept or reject it."[1] Chapter 2, "Science Is Arguing," addressed the term loosely, using critical analysis as the process by which a scientific argument takes place. Critical thinking is the process behind an argument. The main determinant of critical thinking is the ability to judge a claim and change one's reasoning about it if deemed necessary.

It is natural to resist critical thinking; humans are easily taken in by the loud and the superficial. Manipulation by authority and persuasion by a push for conformity are examples of integrity breaches in the scientific process. However, this chapter addresses a variety of strategies to critically evaluate arguments and thereby prevent such manipulation.

It is difficult for humans to engage in critical thinking, and Wallace Stegner best describes this human proclivity in the following statement:

> One of the most difficult operations for imperfect mortals is the making of distinctions, of stopping opinion and belief part way, of accepting qualified ideas. But the individual who can modify or correct beliefs molded by personal interest or the influence of his rearing is rare . . . It is easy to be wise in retrospect, uncommonly difficult in the event.[2]

Critical thinking in science expands the mind beyond its basic confines and creates a scientific way of thinking. Scientific critical thinking transcends the biases of the group and prevents integrity issues that cloud scientific progress. It is in the schools that progress can best be made institutionally to improve critical-thinking skills.

Critical thinking is essential in solving real-world problems. While practicing strategies in school classrooms is a helpful precursor to developing reasoning abilities, real-world challenges require some complex problem solving. Actual phenomena are very difficult to study, so critical analysis often requires a model to be formed about the way something works. As stated in earlier chapters, a scientific model is a simulation of the real world. A model may derive from an investigation or series of studies, but at the core, it requires critical-thinking strategies to solve real-life problems. Models make predictions about how something will work by extrapolating from the artificial condition to the real world. Consider the high rate of esophageal cancer within a population in a province in China. For many years, its cause was unknown and considered a curse upon the people.

While cancer is a natural phenomenon (the abnormal and uncontrollable production of cells forming growths called tumors), its etiology, or cause, is often unknown, leading to all kinds of speculations. For the past two thousand years, a group of people in Lin Xian, China, about 250 miles south of Beijing, have been dying at high rates (one in four) from cancer of the esophagus. Scientists investigated the Lin Xian area to discover that the food being grown by the people was low in the natural chemical molybdenum. This is usually necessary for plant growth, but in its diminished amounts, crops concentrate nitrates from the soil to make up for the low molybdenum levels. Crops also have reduced vitamin C production under such conditions. Nitrates in the plants were being converted to nitrites in the stomachs of Lin Xian residents, and nitrites are linked to various digestive cancers (stomach, colon), including esophageal. The investigators used their critical-thinking strategies along with data to guide them to these discoveries about the local crops in Lin Xian.

It is also known that low levels of vitamin C promote the conversion of nitrates to nitrites. When scientists gave people vitamin C tablets, their production of nitrites dropped. Also, to help reverse the crop production of nitrates, the villagers now add molybdenum to their corn and wheat crops. As a result, nitrate levels in vegetables have dropped 40 percent and vitamin C levels have risen 25 percent. The linked relationships to the villagers' cancer problem are an example of a model, based on scientific investigation and critical thinking, that reveals the truth about a phenomenon without actually manipulating variables. It is a model for the workings of the Lin Xian esophageal cancer phenomenon. It only makes predictions for how the cancer develops in this community based on the data and model derived. However, it is too early to tell whether or not cancer rates will actually change.[3]

INSTRUCTION FOR CRITICAL-THINKING DEVELOPMENT AND THE PRESENTATION OF CONTENT

As emphasized by the variety of state-level guidelines for elementary, secondary, and postsecondary science instruction described in the previous chapter, it is at the college level that schools should tap higher-order critical-thinking patterns. This will limit the often overemphasized superficial coverage of many science topics. The improvement of reasoning is especially advocated in the secondary level standards (e.g., Benchmarks for Science Literacy, 1993, and National Science Education Standards, 1996) to help the transition to college science.

When students are asked merely to memorize information, the process of reasoning development in the curriculum is disregarded. A point of this chapter (which follows in the next section on instructional strategies to improve reasoning) is that college science should emphasize reasoning development. The next section shows that content does not need to suffer when the emphasis changes from content acquisition to critical application. A shift in the focus of undergraduate science expectations is thus recommended.

Although there is no universal instruction strategy to accomplish critical thinking in science classrooms, research shows that some general principles that deviate from current college science instructional practices apply:

- Teach scientific ways of thinking.
- Actively involve students in their own learning.
- Help students to develop a conceptual framework to enhance collateral learning.
- Promote discussion and group activities.

Undergraduate science instruction centered on teacher presentations is not supported by the research. Evidence shows that student grades in large introductory college science courses do not correlate with the lecturing skills or the experience of the instructor. Studies from a number of disciplines suggest that oral presentations to large groups of students increase passivity and contribute little to real learning. Consider that we have had over ninety thousand years of evolution developing all of our senses: auditory, visual, touch, taste, and smell. Teaching to all of the senses makes sense. Despite their limitations, traditional lectures are the most common form of instruction in introductory college science courses. While I would recommend a change from the lecture-oriented instruction, it may be practically unavoidable given institutional policies on limiting instructors while maintaining high introductory course enrollments. A look to the literature on teaching and learning that contains instructional strategies to enhance student learning in lecture settings is beneficial. A synthesis of the research from many studies offers a guide for teaching large science lectures[4]:

- Use paradoxes and apparent contradictions to engage students.
- Make connections with other courses and everyday phenomena.
- Begin each class with something familiar to students.
- Practice your delivery to affect student motivation (e.g., eye contact; enthusiasm).
- Ask divergent (open-ended) over convergent (single-answer) questions.

As noted in the final point, the type of questions asked are important to the kinds of reasoning processes students are encouraged to use. For example, divergent-type questioning such as "Why do birds produce uric acid as an excretory product?" would elicit a much more reasoned answer than the convergent "What do birds produce as an excretory product?" Consider the first example. A student must elaborate on their prior knowledge that uric acid is a precipitate (from chemistry class), that urea produced by humans would kill the bird embryo since it is soluble in the water within the shell (concepts from excretion in introductory biology), and that evolutionarily this is beneficial (a societal and historical principle). The reasoning moves beyond the mere acquisition of the fact that birds make uric acid and enhances the kind of collateral thinking that the state, faculty, and national standards appear to emphasize for postsecondary science learning expectations.[5] Current instructional practices in college science courses use questioning that is convergent. What a shame to lose the interesting questions that drive motivation and further learning.

The improvement of college science laboratory instruction has also become a priority for many institutions. Science laboratories are particularly suited to the development of critical thinking and quantitative thinking and can help students deepen their understanding of science concepts. The incorporation of the following instructional strategies to enhance these expectations is recommended[6]:

- Apply concepts learned to new situations outside of the course.
- Develop experimental and data-analysis skills.
- Experience scientific phenomena as mapping sweat glands on their own bodies.
- Test important laws and rules such as osmosis and diffusion.

Students have differing learning styles. This style determines how they interact with and respond to their learning environment. Thus, teaching strategies helpful for some students may be ineffective for others. Some students learn from lectures most effectively (auditory), while others prefer reading the same material (visual). Instructional strategies that create classrooms providing many opportunities to learn in many ways is best for improving student motivation in science. To illustrate, a genetics laboratory might include materials for building a DNA (deoxyribonicleic acid) model (hands-on), a phone cord (to show structure), a sample of actual DNA (real-life applications), and audio clips from James Watson and Francis Crick on their discovery of DNA's double helix structure (audio). Through the use of a variety of instructional strategies that tap into the multiple intelligences of students, critical thinking and science learning should improve.

THE IMPORTANCE OF INQUIRY-BASED INSTRUCTION TO IMPROVE CRITICAL THINKING

A main learning expectation for every college course should be the enhancement of higher-level thinking abilities in students. Although stressed as an important goal by state-level postsecondary curriculum planners and instructors in the sciences, the majority of undergraduates lack advanced reasoning patterns that are necessary for meeting those learning expectations.[7]

The importance of the development of an inquiry-based, collaborative instructional strategy to improve scientific literacy and reasoning ability is stressed by the Benchmarks for Science Literacy and the National Science Education Standards (NSES) for K–12 science education. A part of scientific literacy is an ability to engage in scientific inquiry, defined by the NSES as the systematic investigation to describe relationships between objects and events through observing and questioning to gather evidence, by comparing evidence, and by generalizing and making inferences to draw conclusions. An emphasis on this aspect of scientific literacy should be a focus of college science programs.

Although college science faculty purportedly advocate instructional methods that improve student scientific reasoning skills, limited research and change in such science teaching has been documented. This can be attributed to a variety of factors. First, college instructors are often scientists who are untrained in instructional theory and practice. As a result, these instructors rely on the methods by which they were taught in order to develop a conceptual framework to guide their teaching.

This framework is most often a traditional pedagogy, characterized by a rigorous adherence to content transmission and not to the development of reasoning skills. Introductory college science courses tend to have large lecture classes that reinforce passive roles for learners. As a result, a special challenge exists to promote reasoning. Also, the undergraduate science laboratories tend to be fact-laden and noninquiry-based, with activities that act in opposition to the development of critical-thinking skills.

Second, there is a fear among many college science educators that content knowledge acquisition would suffer if time were dedicated specifically to reasoning skills. This has fomented, among science educators, a spirit of antagonism against nontraditional instructional methods that advocate critical-thinking development. It is also presumed by these instructors that college students, as adults, should be able to use scientific reasoning strategies independently after reading course materials and listening to lecture presentations. When students are unable to do this, blame is simply placed on deficiencies in secondary level preparation.

Many first-year college students lack the advanced reasoning patterns needed to effectively engage in college science. Several researchers have found that these entering college students are dualistic (right versus wrong only) thinkers who are unable to evaluate an argument based on the strength of the evidence. A number of studies of empirical research outline the deleterious effects of a lack of reasoning ability on achievement in introductory college science courses.[8]

It is thus documented that the ability to judge knowledge claims is critical in understanding science. Traditional teaching methods, such as lecture and textbook assignments, alone are not effective in developing reasoning in students. In the traditional lecture-based classroom, Jean Piaget argues, the teacher is the source of all morality and truth, and "from the intellectual point of view, . . . [the student] accepts all affirmations issuing from the teacher as unquestionable," so that the words are dispensed without the need for student reflection. Thus, a static, unchanging, and factually based way of knowing is perpetuated. This traditional method of instruction consolidates the egocentrisms found in childhood by simply replacing "a belief in self with a belief based on authority, instead of leading the way toward the reflection and the critical discussion that help to constitute reason and that can only be developed by cooperation and genuine intellectual exchange" to improve reasoning.[9] Thus, a major purpose of this chapter is to explore the empirical research on students to determine the truth of the above claims.

THEORETICAL MODELS FOR REASONING

Most of the recent research on the teaching and classification of critical thinking in science incorporates the Piagetian theory of intellectual development, as discussed in chapter 2, "Science Is Arguing." Piaget's model of thought development identifies lower-level reasoning (called concrete reasoning) as being limited to merely the describing and ordering of observable phenomena. The concrete reasoner needs to reference familiar situations to accommodate and assimilate new information. Thus, only an inductive method of analysis (defined as reasoning from particular facts or situations to general conclusions) is used to form conclusions. This type of reasoner lacks awareness of his or her own thinking patterns, and, when faced with inconsistencies in evidence, is unable to generate or consider alternate hypotheses and thus relies primarily on authority and intuition to draw conclusions. In the science laboratory, for example, this student is in need of step-by-step instructions during lengthy procedures.

The higher-level reasoner (called formal reasoner), in contrast, is characterized by the ability to generate and test alternative explanations when confronted with ambiguity. Reasoning begins by imagining possibilities, so the hypothetico-deductive method (defined as reasoning from testing of possibilities) is used to draw conclusions. These reasoners demonstrate the use of formal reasoning patterns, which, for the purpose of this chapter, are defined as the ability to control variables and use of the probabilistic, proportional, correlational, and combinatorial reasoning. This stage also involves the systematic consideration of alternate hypotheses abstract— and evidence to draw conclusions, which is defined as informal reasoning. With such reasoning, individuals can evaluate inconsistencies in their own arguments. Such a reasoner is an independent thinker and can, for example, develop a workable plan of analysis in a science laboratory.

Development of reasoning is related to the individual's ability to understand the nature and defense of one's own knowledge claims. William Perry, an educational researcher, developed in 1970 an epistemologically based developmental scheme exploring how college students make meaning of their educational experiences. He was the first to suggest that reasoning in undergraduates was related to epistemologic maturation. The initial work by Perry on primarily white male college students has led many researchers to explore reasoning in education. Perry's two longitudinal 1970 studies of undergraduates began in the early 1950s at Harvard's Bureau of Study Counsel, and led to a developmental scheme that shows students undergoing epistemological growth, in stages, that results in a maturation of reasoning ability.

During the initial periods of development, according to Perry's model, students view knowledge and produce arguments in a dualistic manner, with right and wrong as absolute and ultimately determined by authority. Thus, in the science classroom, such individuals expect instructors to distribute information without ambiguities.

The progression of student reasoning abilities should continue, according to Perry, through a series of stages characterized by more pluralistic views where knowledge and values are perceived as relative. Perry defined these stages by the level of student possession of higher-level reasoning strategies that use skills to interpret evidence to form conclusions. Thus, the student at this level accepts the existence of possibly conflicting, multiple viewpoints and evaluates the evidence, internal consistency, and coherence of each perspective to formulate a conclusion (called relativism).

According to this model of intellectual development, higher levels of reasoning involve student perception of knowledge and values as contextual and relativistic. Thus, in the science classroom, this informal reasoning translates into skills in interpreting data and observations, evaluating equally valid arguments, and drawing conclusions from experiments. Dualistic, lower-level reasoners are uncomfortable with the uncertainties involved in interpretation and evaluation of scientific evidence, so decision making in science becomes an incomprehensible process when the "right answer" is not provided.[10] Thus, science instruction should help students to relate scientific evidence with conclusions rather than simply focusing on memorization of those conclusions.

Comparisons can be made between Perry's model and Piaget's theory of intellectual development, as shown in table 5.1 (see page 85). The progression of complexity of thought through the various stages is propelled by the interaction of the individual with an environment. Thus, the assumption is that inquiry-based, active learning best develops critical thinking, as advocated by both Perry and Piaget.

Although Perry's scheme addresses general thought development, Patricia King and Karen Strohm Kitchener point out that some aspects of scientific reasoning are not adequately described. Thus, as an extension of Perry's work, King and Kitchener proposed a model that represents the most recent and extensive work on the development of reasoning in college students. The scheme is particularly valuable due to its elaboration on Perry's upper levels of reasoning.

King and Kitchener conducted a fifteen-year interview-based study involving the analysis of reasoning in subjects' responses to ill-structured questions (questions with the possibility of more than one acceptable answer). Through this research, King and Kitchener proposed a seven-stage scheme for reasoning development called the Reflective Judgment model that focuses on the individual's understanding of the nature of knowledge and the process of reflecting on and justifying that knowledge. In fact, my typology of argumentation is based on research findings by King and Kitchener. While my typology focuses on argumentative thinking, Reflective Judgment is derived from responses to ill-structured questions.

There are three levels within the seven-stage model: prereflective (Stages 1, 2, and 3), quasi reflective (Stages 4 and 5), and reflective (Stages 6 and 7). In the prereflective stages, what is observed or what authority dictates determines truth. As with Perry's dualism, the individual is unable to reflect upon uncertainties in answering an ill-structured question. During the quasi reflective stages, there is a growing recognition that the individual cannot know with certainty and that each person is entitled to an opinion. It is during these stages that the belief that knowledge is relative emerges, yet the ability to actively construct arguments and evaluate scientific evidence is absent.[11]

Only at the reflective stages does the role of the knower move from that of a spectator and receiver of knowledge to an active constructor of meaning. Knowledge is recognized as uncertain and relative, so conclusions made from ill-structured questions include the critical evaluation of different positions. The highest level of reasoning occurs (in science) at this stage, when the use of critical inquiry and hypothetical justifications allows for the evaluation and reevaluation of evidence and conclusions for ill-structured questions.

King and Kitchener feel that reasoning abilities develop by assimilating and accommodating existing thinking with new thinking. The mechanics of this model of change are thus Piagetian. A major reason for this section is to determine which instructional variables of the learning environment link the two together.

The higher-level reflective judgment characterizing Stages 6 and 7 has been observed in only a tiny fraction of undergraduates. In addition, although it appears that education is positively correlated with reasoning stages, little development actually takes place during the college years, with less than half a stage during the entire four-year undergraduate experience.

Thus, the studies will show what kinds of teaching methods and instructional environments foster the development of reasoning in college students. Particularly,

Table 5.1. A Comparison of Models of Reasoning through Late Adolescence

Reasoning Level	Piaget	Perry	King and Kitchener
Low	Concrete	Dualism	Prereflective
Medium	Transitional	Multiplicity	Quasi Reflective
High	Formal	Relativism	Reflective

"What instructional methods/environments help self-reflection?" and "Does course content achievement suffer when such methods are used?"

Reasoning is also separated into two constructs: formal reasoning that includes control of variables and correlational, probabilistic, proportional, and combinatorial reasoning; and informal reasoning that includes the ability to explore nature, raise questions, generate multiple working hypotheses, and evaluate evidence to develop a logical argument. A comparison of these models is given in table 5.1.

REVIEW OF RESEARCH ON
COLLEGE SCIENCE REASONING INSTRUCTION

The attempt to change instructional methods in undergraduate science courses to include the development of reasoning is not a recent phenomenon. The earliest study found, by Barnard in 1942, emphasized the need for students to learn more than just factual content.[12] The reform efforts stimulated by "A Nation at Risk" to improve science reasoning have produced most of the studies on undergraduate science found in this section. All of these include a quantitative, experimental design that examines instructional methods for increasing critical thinking in science, particularly biology.

Barnard used a problem-solving method of instruction that emphasized student involvement in the collection of data, the forming of generalizations, and the evaluation of explanations in science over the lecture method. An experimental control group design was used with three batteries of tests administered as dependent variables. The problem-solving group had higher test scores on problem solving than the control group. The author assumes equivalence of the groups based on pretesting and psychological testing and describes the differences in instructional methods in great detail.

However, the results are not convincing due to a number of weaknesses in the study. A modern statistical analysis of variance (ANOVA) should have been done to determine statistical significance of the differences in testing the reasoning development of the subjects in the two groups. It is also doubtful that the subjects of over fifty years ago resemble modern undergraduates. In addition, little information is given about the subjects other than their class years, thus again restricting generalizability.

TEACHING CRITICAL THINKING IN SCIENCE

A theme emerging from an analysis of the studies in this section is the use of writing during instruction to develop student critical-thinking skills. Many educational researchers showed that integrating writing as an expression of reasoning during instruction has a positive impact on student reasoning development.[13]

The use of a "Science News Exercises" instructional method in introductory college biology with a pretest/posttest experimental control group design by Robin Tyser and William Cerbin showed improvement in student reasoning skills. This method presented students with a model for evaluating evidence in popular science articles to develop a logically persuasive argument. Students assessed six to seven biweekly scientific articles using guidelines for the direct teaching of reasoning through a three-step line-of-reasoning model. This is the only study found to directly teach and apply a method for informal reasoning. The model gives simple guidelines for the identification and evaluation of evidence and for then persuasively communicating a developed article. The Science News Exercises group performed statistically significantly better than the traditional lecture group on the objective test for evaluating evidence and on the line-of-reasoning written test. Content achievement was not assessed, but the authors contend that only two hundred minutes of lecture time (10 percent of the lecture course) were used for Science News Exercises. Thus, the concern for a loss of content should be ameliorated, according to the authors.

Although Tyser and Cerbin used statistical analyses to compare the means, several weaknesses are evident that cast a doubt on the results. First, there is little subject information offered except that 80 percent are nonmajors. This limits generalizability, especially to courses with a high proportion of biology majors. Second, the teachers for control and experimental treatments differed, thus introducing the possibility of extraneous variables. Third, the authors do not offer reliability and validity of both the objective and written tests. Despite these flaws, the results do show evidence of positive effects of nontraditional instruction emphasizing writing on reasoning development in college students.

A study by Anton Lawson and Donald Snitgen on the direct teaching of formal reasoning in an inquiry-based course for preservice elementary teachers also showed positive effects on reasoning development. The course, titled Biological Science for the Elementary Teacher, used reasoning modules to facilitate collaboration among students to apply formal reasoning strategies to experimentation. This is the only study in the section to address the direct teaching of formal reasoning. The authors implement Piaget's suggestion to ground the development of formal reasoning in concrete experiences and social interactions. Their method introduces what is familiar to the student, and, through collaboration, allows the students to recognize their own faulty reasoning.

This Piagetian model was pretested and posttested using an experimental design, but it lacked a control group. The authors report statistically significant pretest and posttest increases in formal reasoning for the subjects after taking the course. The

Lawson Test for Formal Reasoning was used and verified for validity and for reliability, according to the authors.

However, the test for the transfer of reasoning to unfamiliar contexts showed no significant improvement among subjects. Thus, the *application* of reasoning using this method is not demonstrated. In fact, qualitative analysis of the results indicated that students misapplied reasoning strategies even though they could formally reason. A future study should investigate possible negative effects of this method, such as confusing established formal reasoning patterns.

Other qualitative data obtained by this study were particularly illuminating. Some formal reasoners found the course "childish and boring," and others dropped the course, citing their desire not to conform to the thinking methods called for by the Reasoning Modules. Although the authors cite positive comments for the course on reasoning development, no qualitative analyses were done to draw definitive conclusions. Overall there is a lack of generalizability to a major course since all subjects were nonmajors and an absence of a comparative group is a flaw in the research design. Thus, although ostensibly demonstrating positive effects of this writing-based inquiry course, it is not prudent to draw definitive conclusions until the flaws in the study are corrected and the study is repeated.

Michael Moll and Robert Allen used an inquiry method of instruction using video presentations to guide discussions to develop reasoning skills in response to ill-structured questions. This course emphasized student writing to create arguments from an analysis of evidence to develop informal reasoning. The method used was described in detail by the authors as stressing student exploration of ideas, interpretations, and various lines of informal reasoning to improve critical-thinking skills. These skills are identified as informal reasoning, as defined earlier.

To show a significantly higher improvement of reasoning skill and content knowledge ($p < 0.001$) by the experimental group, an experimental, no-control group design was used. The gains were not shown to be related to gender or major. The authors also cite qualitative evidence that students appear to reason better after taking the course. The weaknesses of the study again cast doubt on definitive conclusions. The lack of a control does not allow for isolation of the effects of maturation of reasoning over the semester, no reliability is mentioned for the tests, the statistical methods used are not given, the number and description of subjects are omitted, and the types of qualitative methods used are not discussed (for example, questionnaire and survey).

If this information were given, the study would be particularly interesting since it is the only one to explore the interaction of gender and major with instruction to influence reasoning. In addition, since one group in the study was given more content and scored significantly higher on the content posttest but not on the reasoning posttest than the other groups, this implies that content alone was not sufficient for improving reasoning.

Diane Ebert-May, Carol Brewer, and Sylvester Allred conducted the final inquiry-based approach emphasizing writing to develop reasoning skills. Care was taken to control variables in instruction in this pretest/posttest quasi-experimental control group design. The experimental lectures in nonmajors introductory college biology were based on a modified learning cycle of the biological sciences curriculum study (BSCS) model of instruction, in which there was a high level of student involvement and a risk-free atmosphere to facilitate student collaboration in constructing answers to biological questions. The writing assessment included one-page papers and group work to answer ill-structured questions. The comparison lectures were traditional and factually based.

Results from an analysis of variance (ANOVA) indicated that students in the experimental groups scored significantly higher on process/reasoning questions (identified as informal reasoning for the purposes of this study) on an NABT (National Association of Biology Teachers) exam. Also, in support of the view that such nontraditional inquiry-based teaching does not negatively affect content achievement, no significant differences were found between the groups in terms of content questions on the NABT exam. Ebert-May et al. contend that, in the end, the amount of material covered in the activity-based classroom is equal to the material covered in the traditional classroom. Considering the importance of content coverage for student progression to established professional programs in science (i.e., medicine, dentistry), future studies should replicate such a design, paying particular attention to those standards set forth by preprofessional advisory committees, professional school entrance exams, and professional school admissions.

Qualitative data obtained through random selection of students for interviews and written responses indicated that students were changing the way they viewed the acquisition of knowledge. "Students began questioning the nature of the scientific evidence before them" and "were more likely to apply their understanding of biological concepts to personal, public, and ethical issues than if they had experienced the traditional lecture format," showing the development of informal reasoning as defined in this review. A well-constructed qualitative analysis such as this can reveal information that quantitative designs cannot.

Thus, the research design by Ebert-May et al. represents the strongest evidence presented to date in support of inquiry-based, collaborative instruction as a means of improving reasoning and not weakening content acquisition. It was a mixed-method approach that used both quantitative and qualitative techniques that together allowed for a broad exploration of the variables. Unlike the previous studies described, this research design includes comparison groups, control of instructional variables (e.g., same lecture notes and instructor), the statistical analyses mentioned, a heterogeneous, large sample size (559 subjects), and an appropriate qualitative methodology.

There are, however, some unanswered questions remaining with regard to testing (e.g., the reliability and validity of the NABT exam are not given). This is not problematic if it is assumed that such a national exam has sufficient reliability and validity. However, it is a high school exam, so prior contest achievement not related to this college course, which is not addressed, could have influenced the results. For a future study, an exam more appropriately measuring college biology content achievement should be implemented.

The final study supporting the view that inquiry-based learning improves student reasoning used a pretest/posttest control group quasi-experimental design by Gerry Haukoos and John Penick.[14] It is the only study in this section that treats the community college level. The effects of a discovery classroom climate (DCC) were compared to a non-discovery classroom climate (NDCC) in terms of student achievement in biology and the learning of reasoning skills. There were seventy-eight subjects divided into two sections of ten-week-long NDCC courses, one section of a ten-week DCC, and one section of a five-week DCC course.

The authors describe the classroom climates in detail. In general, the differences were based on the directness of teaching. In the DCC, teaching is indirect, with content dialogued and discovered through ill-structured questioning. Thus, students construct the knowledge, as in the other studies shown so far. In contrast, the NDCC was the traditional lecture similar to the comparison methods seen in Barnard and Ebert-May et al. An ANOVA showed that students in the ten-week DCC group scored significantly higher on the reasoning skills exam ($p < .01$) as compared with the other groups. There were also no significant differences found between the groups in terms of the learning of biological content. Haukoos and Penick are the only researchers in this section to explore the interaction of time and instruction on reasoning and achievement. Since the five-week DCC does not show significantly improved reasoning as compared with the ten-week DCC, this implies that enough time must be available to develop critical thinking in science.

For each of the aforementioned studies in this section, it would be interesting to explore the relative contributions of different variables within the instruction that led to the successful development of reasoning by the authors' methods. For example, although all of the studies used both collaborative and inquiry methodologies, what were the relative contributions of each of these variables to elicit change? If a noncollaborative approach were used, how would the results on reasoning development change, for example? Also, how would the introduction of a more intimidating, yet nontraditional classroom environment that harms positive attitudes, change the results?

This raises an important point—is the teaching of reasoning even possible? Students could be intrinsically locked into a Piagetian developmental stage of reasoning until chronologically matured. The ability to change their predisposed abilities before natural development allows may not be possible. It is the contention of this section that instruction can affect reasoning ability, but the evidence given by the five studies favoring this view does not address the mechanism of change in reasoning—it remains a black box. Thus, although empirical results show increases in reasoning levels through the instruction suggested, no specific instructional variables are explained as to why they are causing change. The variable interactions are given in figure 5.1.

It is thus seen that the studies discussed in this chapter would allow for stronger conclusions to be drawn if repeated with their respective weaknesses ameliorated.

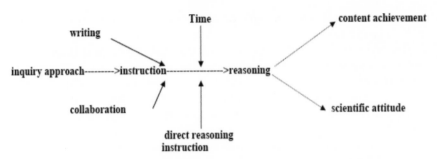

Figure 5.1. Interaction of instructional variables to improve science reasoning.

Also, the many questions that arise when considering the studies more critically show the gaps in explaining what and how different instructional variables play a role in changing reasoning. However, it is clear that the successful studies affecting reasoning improvement cited in this section, when taken together, strongly support the use of nontraditional, inquiry-based, collaborative methodologies for the development of student reasoning.

GENERAL RECOMMENDATIONS
FOR POSTSECONDARY SCIENCE

The following statements can be made with strong support from the empirical research of the studies reviewed, although they are not submitted without contestation. Several weaknesses were found in some supporting studies, and there is not unanimous agreement on all points in the following recommendations for science education.

1. Inquiry-based, nontraditional, collaborative instruction is more effective than traditional, lecture instruction in developing higher-order critical-thinking skills in science courses.
2. The gains in reasoning through inquiry-based, nontraditional, collaborative instruction are not achieved at the loss of content acquisition.
3. Inquiry-based, nontraditional, collaborative instruction emphasizing writing to develop reasoning has higher success at developing a student's critical thinking than those methods not emphasizing writing.
4. The direct instruction of formal and informal reasoning leads to gains in those reasoning skill areas.
5. Gender and major do not appear to interact with instruction to influence reasoning.
6. Enough instructional time is needed to improve reasoning.

7. Inquiry-based instruction that improves reasoning also enhances positive scientific attitudes.
8. The laboratory is an important part of the science curriculum because it improves critical-thinking skills when inquiry-based and collaborative.

CRITICAL-THINKING STRATEGIES

Critical thinking is a task that requires constant practice, and no person is a superb thinker under all circumstances. There are the factors of human emotion, time needed to make a decision, and complete or incomplete information available on any given situation. I was recently debating whether or not to make a large purchase, evaluating all evidence and counterarguments, and yet I was distracted throughout the process, making a logical decision impossible.

Clearly this is an extreme example of challenges to scientific thought. However, Richard Paul developed a list of strategies to help best generally meet this ideal.[15] He divided his advice into two categories: an *affective* set of strategies and a *cognitive* set of strategies.[16] Affective measures refer to those involving feelings, emotions, and attitudes. Cognitive strategies involve brain or intellectual skills, such as interrelating ideas, memorizing, and developing hypotheses. In other words, these strategies incorporate mental processes. Each of Paul's strategies (shown italicized in the next sections) are discussed in the context of science reasoning development in everyday life.

AFFECTIVE STRATEGIES

An ability to see the world from the outside in is described by Paul as both *thinking independently*, or "outside the box," and developing insight into *egocentricity* or *sociocentricity*. By that Paul refers to not becoming caught up in one's own "little world" and instead viewing ideas through the eyes of others. This is discussed in the Bible quite frequently as a theme of not being so judgmental.

There is a short story titled "Body Ritual of the Nacirema" by Horace Miner. The story portrays people of a "tribe," described as very savage to both their own bodies and each other. The readers are gently led to the different institutions of the Nacirema society, each depicted as practicing base and abhorrent rituals. At the finale, readers might judge the Nacerima harshly. The twist to the story comes only after seeing a parallel (after much critical thought) between U.S. society and the Nacerima. Of course, "Nacerima" is really "American" spelled backward. This exercise is a strategy that helped me see my own egocentrisms, and it broke my inability to openly criticize our society as others would do from the outside. Very often, growing up in Queens, New York, I encountered many cultures, and only through getting to know other people did I break through my sociocentric tendencies.

Good scientists require an ability to view their research questions and designs from another perspective. Science requires that there is an objective reality and that one should think about his or her own biases, both personally and societally, to overcome them. In this way, new insights and discoveries are made in sciences that transcend individual bias.

In the show *The Big Bang Theory*, scientists are depicted as emotionless and valuing only the objective. The show is funny for many reasons, but often, their stereotypes are far from the truth. The truth is that scientists are humans with feelings. Paul suggests *exploring thought and underlying feelings* about objective data to see the connection. For instance, a medical doctor may really like a patient personally and somehow overlook serious symptoms because he or she wishes that the patient was not gravely ill. At times, the medical community must practice removing themselves emotionally from a physical exam to evaluate objectively. This requires not only practice but also a tremendous amount of self-awareness.

Affective strategies also involve an evaluation of one's confidence levels. On one hand, Paul suggests *developing intellectual humility*, and on the other hand, *developing confidence and courage in reasoning skills*. For example, a humble person will not get caught up in the power of his or her own reasoning and is more likely to see all sides. That said, confidence is gained by advances made with objectivity. Nonconformity and standing up to political or other social forces to uphold objectivity is its own reward. Paul's final affective strategy, to *develop intellectual good faith or integrity*, will prevail in the end. The ontological result will be the advancement of science. However, these goals are often difficult to reach because society and its influences get in the way of integrity and objectivity.

COGNITIVE STRATEGIES

Scientists deal with reality in simplified forms and produce models and experiments that often only simulate the real world. Natural phenomena are actually very complex, and scientists know this. However, there is compromise to studying real-world phenomena. That said, scientists must *avoid oversimplifications* or *generalizations* and make sure to study the complexity of the larger situation. Biology of the kidney can be easily taught by listing the functions of the kidney. Students then view this representation as a discreet representation of its physiology or anatomy. In reality, there is more going on than even science can figure out. There are also more complex interactions with other organs than a simple list given to students can realistically cover.

Instead, Paul argues that *analogous situations should be compared to give insights to new contexts*. In this way, existing ideas can be explored (ideas about the kidney, for example) and expanded or applied to new situations to further theories and pattern statements. For example, when presenting kidney physiology as a subject to be taught, links to hypertension and effects on kidney cell structure make comparisons that further thinking about content.

In order to develop *one's perspective by creating or exploring one's beliefs, arguments, or theories*, a few cognitive strategies should be used. First, a clear statement of a *thesis* should be sought to ensure that the investigation is well defined and well understood. Second, *standards for an evaluation* should be developed to make sure any investigation's results are objective and standardized. Third, the reliability of the *sources* for data should be thoroughly evaluated because many sources are quite biased or simply inaccurate. Many sources do not fit the parameters of a study or are questionable . . . for example, does one take the barber's word for it that they need a haircut? My barber is actually quite honest! Sources should also be scrutinized for validity (how well it measures what it is supposed to measure) and reliability (how consistent the results are). Fourth, *analyzing or evaluating arguments, interpretations, beliefs, or theories* cover a set of strategies. Perhaps the most important aspect is to evaluate counterarguments or alternate viewpoints to one's beliefs. The best critical thinkers are able to consider the points of the arguments against their own. It is important for true critical thinkers not to simply overlook or dismiss the tenets of the counterargument. Objectivity requires that a situation be considered seriously until disproven or falsified, as discussed by Karl Popper in chapter 2.

Good critical-thinking skills require reasoning *argumentatively*, as discussed in chapter 2. Thinking to oneself and speaking to others to get at as many perspectives as possible is an excellent strategy critically reflecting on scientific phenomena. Through this process, *dialectic* thinking is practiced, in which conflicting ideas are set against each other to determine the strengths and weaknesses of each of the ideas. To accomplish this, the *Socratic* method of discussion is needed, which involves asking the right questions to prompt discussion of the strengths and weaknesses of any argument.

These methods allow for another of Paul's strategies, to *distinguish relevant from irrelevant facts*. A scientific argument weakens as more irrelevant data obscure it. This requires strategies to think about one's own thinking, called metacognition. Paul states that "one possible definition of critical thinking is the art of thinking about your thinking while you are thinking in order to make your thinking better: clearer, more accurate."[17] It is clear that the use of certain vocabulary should be used during metacognitive thought: *conclude, assume, bias, relevance, evidence, justify*, and *consistent* are all useful terms for evaluating scientific claims.

Finally, the importance of *evaluating evidence and the alleged facts* suggests that "not everything offered as evidence should be accepted. Evidence and factual claims should be scrutinized and evaluated. Evidence can be complete or incomplete, acceptable, questionable, or false."[18] No one is a superb critical thinker. We are all taken in by bias, confusion, and society at many times, depending on the issue at hand. The key is to practice critical-thinking skills in the ways mentioned to develop into a more critical thinker. Science requires this of scientists, and society requires this of a scientifically literate populace.

II

SCIENCE IN EVERYDAY LIFE

6

The Media

HOW TO THINK CRITICALLY
IN OUR SOCIETY: FACEBOOK STUDY

The reader is invited to engage in a treatise on how scientists think critically about the results of research reported by the media. The media is defined, for our purposes, as all of the means of mass information transmission available to the public. This chapter describes the role of the media in promoting and many times contaminating scientific thinking. Its main goal is to apply useful strategies in evaluating scientific research to counter the media's often destructive role in educating people about science.

Examples throughout the book thus far exposed how easy it is to make erroneous decisions based on media bias, inaccuracy, and poor methodology and foundational science. The reader becomes savvy to the pitfalls of misinterpreting and misusing science data by practicing the critical-thinking strategies presented in the previous chapter. Studying research presented by the media is a difficult task that requires science literacy. This chapter also suggests educational reform to ameliorate the effects of a changing society centered on media-style learning and science knowledge acquisition.

To illustrate, consider the research discussed in chapter 1, "Introduction," reporting the "discovery" made that "people who have more friends on Facebook have better developed regions of the brain." The reports were based on the abstract (summary) of a research study by the *Proceedings of the Royal Society B: Biological Sciences*, a journal reporting research in neuroscience, as presented in the following textbox.[1] In it, 165 participants' brains were scanned using MRIs, and each subject was asked to report the number of Facebook friends they had. When considering the abstract

---

JOURNAL ABSTRACT REPORTING RESEARCH IN NEUROSCIENCE

Online Social Network Size Is Reflected in Human Brain Structure, by R. Kanai, B. Bahrami, R. Roylance, and G. Rees

The increasing ubiquity of web-based social networking services is a striking feature of modern human society. The degree to which individuals participate in these networks varies substantially for reasons that are unclear. Here, we show a biological basis for such variability by demonstrating that quantitative variation in the number of friends an individual declares on a web-based social networking service reliably predicted grey matter density in the right superior temporal sulcus, left middle temporal gyrus and entorhinal cortex. Such regions have been previously implicated in social perception and associative memory, respectively. We further show that variability in the size of such online friendship networks was significantly correlated with the size of more intimate real-world social groups. However, the brain regions we identified were specifically associated with online social network size, whereas the grey matter density of the amygdala was correlated both with online and real-world social network sizes. Taken together, our findings demonstrate that the size of an individual's online social network is closely linked to focal brain structure implicated in social cognition.

(summary of the investigation) in the box above, use the rules for critical thinking to evaluate the research methodology and mathematics. Use the tools and strategies scientists covered in earlier chapters to determine the relevance of the study.

MEDIA'S ROLE IN CRITICAL EVALUATION

Does using Facebook really make a person more adept at socializing? Immediately after the Facebook study in the textbox above was published, many news reports claimed just this—a headline that Facebook socializing actually "*develops* portions of the brain to gain better skills in social functioning." This misrepresents the actual study shown in the textbox. The media headline reports a causal relationship while the actual research indicates only weak correlational results. Just because a relationship between brain size and social network was discovered in a small number of subjects, it does not mean that Facebook is its cause. Further, there is little information on statistical analysis or study power from either the researchers or the media reports to draw such conclusions. A recurrent theme in media misinterpretation of scientific research is confusing causation with correlation.

Why do media outlets make such claims to the public? A commonly cited explanation is that there is limited space or time available for news organizations to give all of the facts; that if a person is interested in knowing more, she or he can research the topic herself or himself. In the face of the poor science literacy skills reported by international assessment measures described in earlier chapters, it is surprising that little to no effort is made to educate the public.

There must be more to the media's poor reporting of science. After all, sports reports and articles contain far more data (some newspapers have entire sections dedicated to sports) and have more time devoted to them than do science research findings. Reporting by the media of such findings as in the above example of Facebook may show both political and economic agendas to push technology—a kind of groupthink—to save the U.S. economy.

Consider that if the public accepts that Facebook is good for it, then the part of our GDP (gross domestic product) dedicated to Internet commerce and online media exposure will strengthen. Or is it simple incompetence in reporting? Either way, the media acts as both a help and a hindrance to scientific thinking. In this chapter, we will explore the role of science education in filling the gap in science literacy and the problems created by a media-driven culture. Reform efforts are suggested for the culture to help the public become better consumers and users of science research.

THE MEDIA AND SCIENCE EDUCATION: TOO CLOSE FOR COMFORT

A close parallel between a society yearning for entertainment and an educational system willing to deliver that desire has led to a societal dependence on the media for information. While science is done in everyday life, the two main sources of science learning either emanate in traditional schooling or through media outlets. The second source is inadequate due to its fragmented methods of information transmission. Technology is very much used to enhance learning science; for example, through Internet research or through the use of statistical programs in data analysis. However, mass media information, as seen in examples throughout the book, constitutes a poor representation of the scientific method. While the media is indeed serving as a primary source of science knowledge, its instability and inaccuracy require a competitive curriculum in the schools to counter its effects. The most important source of science is our schools.

In times of rapid social and scientific changes such as ours, schools should conserve a stable system of learning. Neil Postman (1931–2003), an educational researcher and media critic, argued that TV functioned as the nation's main curriculum. Much of his work analyzing the media is reflected in this chapter. He defines a curriculum as a specially structured information system whose purpose is to teach, train, and cultivate the minds of youth. The media continually competes with and obliterates the school curriculum, according to his analysis. For example, school curriculum is based on words, whereas TV is based on pictures. Each culture perceives truth as being based on certain symbolic forms. In the Western world, these forms have changed over the centuries.

The main carrier of truth developed from the oral tradition (with oral exams and recitations in schools) to the written (seen as more authentic), and now the visual through media presentation. Oral communication can be elaborate but lacks continuity. For example, a lecture is only heard; it may be complex and engaging, contributing to science, but it is lost after it is presented. Most ancient cultures transmitted information orally. The Babylonians through the Egyptians transferred knowledge in this way, except for some scatterings of writings. The move to written language amplified during the Middle Ages with the development of the printing press. The power of words on paper fomented great changes in society and science. Once ideas were written, they could be expanded upon over generations of scholars. Educational systems were built upon by written information.

The printed page leads one away from the form of its symbols and toward its meaning—which is thought provoking.[2] The word *art*, for example, is interesting and divergent not because it has *a*, *r*, and *t* as symbols, but because the word itself connotes many possible values. The term *art* could elicit an image of a man named Art, perhaps how you knew that person in your life; it could connect to a traditional painting of a scene of an island in Lake Maggiorie, Italy, or a painting that your mother liked; it could also bring back a memory of when you first saw a modern artwork or how you think of art as a way to express political views. However the images manifest, the written word is clearly a divergent entity that allows thought to be complex. Science requires this kind of intricate thought to be best understood and expanded upon.

The shift to a visually based culture of learning impedes this type of mental processing. The TV or Internet image, because it is interesting in and of itself, leads toward the form and away from the meaning. A picture of a businessman named Art ends the story behind the word. This converges thought to a single meaning. Art could have had so many directions, but the image guides the observer toward one main focus. When schools attempt to emulate a visual style of pedagogy, developing curriculum around technology and images for students instead of around written language, a danger to scientific thinking is clear. A curriculum dominated by these concrete, non-thought-provoking visual images weakens the abstract powers of youth. An image of Art is shown in figure 6.1. Unlike the written word, its meaning is convergent and sets the thoughts of the viewer in a particular direction.

Thus, TV and Internet imaging have shifted the values of society so that image is more important than ideas. Consider our former president one hundred years ago (1909–1913), William Howard Taft, who was three hundred pounds and quite unattractive. I posit that he would not have been elected today in this visual age, and we would have thus lost out on his ideas and qualities.[3] Presidents since the TV age have had to maintain the "right" image to become electable. Richard Nixon lost against John F. Kennedy in 1960, in part because of his poor debate appearance and by showing a five o'clock shadow on his face. It is generally accepted that Ronald Reagan, Bill Clinton, George W. Bush, and Barack Obama were all elected in part based upon their appealing visual image.

Figure 6.1.   A man named Art. Source: Hemera/ThinkStock.

While seemingly entertainment-based, the movement to a visual culture is a direct threat to scientific progress. The media's curriculum is in direct opposition to a science curriculum that promotes critical thinking. How does a science-based curriculum differ from a media curriculum? First, skills take time and effort to foster in science programs. Expertise in microscope use requires sequential development and can always be improved upon. Internet and TV watching requires very little skill—one is no better now than after ten years of practice. Second, there are no standards for which to strive to achieve in a media-based curriculum. Schools require testing and levels of competency of their students to measure progress and redress inadequacies. The attempt to implement consistent standards across schools has been the admirable part of the No Child Left Behind Act. While standardized testing is loathed for many reasons, the attempt to measure achievement along standardized lines should be advocated. Third, the goals of the two curricular structures are vastly different: while school goals are content- and skill-driven, the goal of a media curriculum is merely to capture interest. Fourth, schools work with words, as discussed earlier. Words give a concept and idea, which must be analyzed and defended. Media outlets, image-centered, require little to no analysis of their presented ideas. Words are remembered, but visual images are soon forgotten. An image "says it all," but written words in schools require more complex thought and debate. One cannot analyze whether a McDonald's commercial on the Internet is true or false in its claims, and its success is dependent only upon whether or not the representation appeals to the public. Discussion of a commercial's validity can occur in a classroom setting or within a text or written document. Analysis cannot occur within the media's confines of a curriculum. A person is often isolated from collaborative thinking when watching TV or even when on the Internet. While there are chat rooms and ample information available on Internet sites, there is little verification of the reliability or validity of the sources and no guidance by professionals on how to interpret that information.

This has led to an explosion of increased belief in pseudoscience, bad science, and outright lies about science. Information is freely given now as never before, with so little guidance and grounding in a largely scientifically illiterate public. As discussed earlier, the declines in science literacy endanger public health and decision making. All of this stems from an increased reliance on the media curriculum.

A school's science curriculum requires continuity with past learning, with skill development built upon over many years and in many subjects. Writing and mathematics literacy intertwine with science learning because they are required to incorporate new content into existing thought patterns. There is coherence around themes: measurement in chemistry and life functions in biology, for example. Thus, prior learning is needed to build upon for new learning around science concepts. Thus, the fifth difference between the two curricular structures is that the media curriculum is always presented in the present tense, with its form and substance requiring no prior knowledge, while science courses require an adequate background. One is able to simply enter into the visual world of information on the Internet or on TV without ever having to have been a part of that world. While watching a soap opera on TV the other day (perhaps *Days of Our Lives*) to avoid writer's block, I realized that I had never seen the program and yet was completely into it, images and plot, within about thirty seconds. I cannot recall anything that happened on the program, but I was instantly entertained by the events. In fact, the average shot on a TV program is 3.5 seconds. This "instant gratification" style of learning limits the kind of patience required to do scientific investigations.

When a person is accustomed to a fragmented curriculum, it is difficult to shift into a science classroom that requires a delay in gratification to see results. For example, in a laboratory experiment (which is carefully developed to actually show results within a certain period), it takes hours and skills to develop outcomes. Classroom laboratory learning also requires analysis and prior knowledge (e.g., mathematical abilities) to determine answers and actually see the point of the experiment or procedure. Science is not usually proscribed as in a laboratory procedure. Science results, at times, require years or, in some cases, are never obtained. A delayed gratification for answers to questions is integral in successful scientific thinking. Juxtapose this required characteristic of the science student with an average eight-year-old who sees thirty thousand different shots per week on TV and countless images on the Internet.[4] Is this student able to shift into a deferred gratification temperament after years of fragmented media and instant entertainment?

MEDIA-EDUCATIONAL COMPLEX

The supplanting of science learning by media-style pressure is not the fault of the public. People are attracted to easy entertainment, but it is the media that falls to the lowest intellectual levels to obtain ratings. The media are teaching people to accept low levels of programming. It is human nature to want to be entertained and have easy access to instant gratification. The Internet and TV provide such an outlet. But the media capitalize on our lowest element—immediate gratification.

The media choose profit over educational programming by promoting low levels of thinking, which are easy to create. There is an instant audience for cheaply made soap operas, but science programs using scholars in African ecosystems are more expensive to create and have a smaller audience. Obviously, people will watch reality TV before they watch *NOVA* because of the foundational structure in our society. Science literacy (or a lack of it) is at issue; if the public is educated in a way that does not give it access to science (e.g., people lack the abilities to do science because they were not successful in schools), they will not be interested in such programming. If, however, the educational system could turn people on to science, the media would offer more educational programming because it would be in more demand.

The relationship between simple thinking and simple programming is clear. It is the fault of both the educational system and the media. There is a relationship between the two feeding one another in this attempt to keep people isolated from science. The media lure people and the educational system to view knowledge as simple and instant. Science is not simple and rarely instant. Thus, science is not compatible with the current media curriculum and its influence on educational systems. The educational system, afraid of turning students off to science, attempts to bring technology and a media-style learning into the classroom to make it more fun. This is where the danger in the media-educational relationship begins.

Resistance to a temptation to modify curricular structure to mirror a media-style learning approach is vital at this juncture. A society that is very focused, for a host of reasons, on technology entertainment as a solution to U.S. economic problems has begun to exert pressure on the schools to deliver an education paralleling media outlets. There appears to be a development of a media-educational complex, which is defined as an implicit and explicit alliance between the technology sections of the economy with educational institutions.

This alliance is supporting itself to gain control over public opinion and money. The media sell the need for technology to the schools, and schools use technology as an integral part of their instruction. This teaches students a style of education, based on the media, that is fragmented and instant-gratification focused. Instead of studying hard through a series of chapters, students are taught that it is OK to merely skim through PowerPoints. A move from intense study to instant overviews is becoming more and more acceptable. I am not against technology in the classroom, because it is a vital part of learning. However, its emphasis and expense need to be in check. Also, the ways technology uses us should be exposed.

An understanding of technology is vital for our youths to be competitive in the job market and research fields. However, the degree of emphasis on technology and media throughout the school curriculum may be unnecessary. Over thirty-five million people learned to use computers without the use of schools by the year 2000. I learned science with a card catalog, typewriter, calculator, and pencil. I do not advocate going back to those inefficiencies, but most of my generation learned computers on their own, without official schooling.

Thus, intensive training in computers is unnecessary for most students. With the new expenses of technology, the United States spends more per capita on students and yet falls into the lowest categories in international mathematics and science test results. Much spending can be attributed to the emphasis on technology uses in the classroom. Are iPads, PowerPoint, Excel, and Word really needed for a first grader? This increased spending in the past twenty-five years has not increased results in science learning, as demonstrated by national and international test results.

A main motive for the media-educational complex is to get schools to buy technology. The educational system becomes a major consumer (with taxpayer dollars) of technology. The technology sector uses media outlets to educate the public that technology purchases are essential for their children. After all, the news and advertisements devote a disproportionate time to technology news coverage. The latest iPhones and iPads are given far more coverage than science developments or political changes. With the explosion of deficit spending, the case is more and more difficult

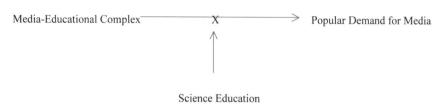

Figure 6.2. Combating the media-educational complex.

to make for the media-educational complex. Cuts in spending throughout the nation have led to a call for greater stewardship. In protest, the media have doubled down and increased their coverage of the positives of the latest technology developments. Figure 6.2 shows the relationship of the media-educational complex in its influence on increasing demand for itself through the schools. The effort by science educators to improve science literacy is noted as a main combatant to this relationship. Through improving science literacy and elevating the populace as a whole, demand for lower levels of entertainment will decline. In this way, scientific thinking and the development of science will extend throughout the culture.

What are some consequences of this pressure on the teachers? Teachers are victims of the drive to incorporate technology into the classroom and emulate media-driven standards. Many teachers are trying to make their classrooms into second-rate TV shows and gaming. There is an increase in visual aids, a decrease in reading assignments, and a decline in writing and science abilities, as demonstrated on TIMMS (Trends in International Mathematics and Science Study), PISA (Programme for International Student Assessment), and other standardized tests. Students are expected to be charmed and amused. We have all heard the slogan "Learning should be fun." It is true that a teacher's goal should be to increase student motivation so that students go on to independently study and become lifelong learners, but science also requires times when learning is based on hard discipline to memorize terms and to work through difficult material to obtain knowledge and insights. Without these events, full understanding in science cannot be achieved.

Schools place pressure on teachers to make learning fun in many ways and for several important reasons. Students are happy in a "fun" classroom, so there are fewer problems with both students and parents. Also, technology sells and is popular with the media's promotion of itself and its related technologies. This pressure is emblematic of an acceptance of the type of student entering into the classroom. The media culture has taught kids to do what's required of them—to be passive, demand gratification instantly, and lose the ability to think critically.

There is much evidence that learning decreases when presented in a dramatic setting. Postman cites studies with student responses to a TV news program indicating that 51 percent could not recall a single item of news a few minutes later. Learning in the media culture occurs within the sensory register of the brain. In educational psychology, learning new information requires that it is received into the sensory register, which is a collection of nerves within the brain that receives information from the outside world—sight, smell, hearing, taste, and touch. However, that information needs to be rehearsed within one's short-term memory to become useful to the brain. For example, remembering a phone number requires rehearsal a few times to keep it within the brain to dial the number. However, its movement into the long-term memory for storage and retrieval is somewhat more involved, as shown in figure 6.3. A phone number needs to be assimilated and accommodated through using strategies to incorporate that phone number. Perhaps 366-1012 is similar to an address you once knew or to your grandmother's phone number. Its unique placement within your brain structure requires time and some amount of placement within you. The building up of neuron pathways is documented in research as evidence that studying and learning increases permanent bridges of knowledge within the brain. Thought processes, when practiced, build up new strength just like a muscle. The hippocampus of the brain is mitotic, meaning that it is able to divide and improve short- and long-term memory. The hippocampus brings short-term memory information into longer-term storage. *50 First Dates*, a movie depicting a person who could not remember any new information, is an example of damage to this processing model.

The media-educational complex drives a style of learning that remains within the sensory register of thought processing. In some ways, it caters to a population just like the character in *50 First Dates*, assuming that we have little or no ability for recall or thought processing. Information is not often moved from short- to long-term memory. Lower levels of observation are required in a visual curriculum, whereas higher levels of critical analysis are cornerstones in science learning. The incompatibility of science education with the media-educational complex has fostered animosity between the two. Media time and depth of coverage for science topics remains limited, and science images are pejorative. This is not a conscious effort by the media, in my opinion, because it results as a natural outcome to an observation-based and scientifically deficient society. In the media-driven education, meaning is secured from TV and Internet sources in segmented, nonsequential, and noninferential ways. Learning is not collateral, so it cannot be remembered long term. The media only service this kind of a public, which they have created.

Interest in learning science is highest during childhood years. Studies show that most students have an intrinsic motivation to learn science and answer scientific questions during the elementary years. A marked decline in interest occurs at every

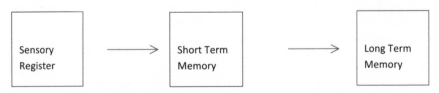

Figure 6.3. Information processing model.

grade level and at every interface between institutional and degree advancement in the sciences. There are many reasons for this, as will be addressed in later chapters, but harnessing the energy of youthful inquisition is essential to recruiting new talent in the sciences.

The media curriculum has put an abrupt end to childhood and to the natural curiosity that drives science in youth. Media programming is geared mainly toward twenty-somethings in the United States, so it is producing a society in which children are trying to act like adults and older people are persuaded to long for their youth. This is apparent in the increased demand for cosmetic surgeries within the past two decades, for example. At the same time, youth is expected to conform to fashion styles and speak in ways that "fit in" with the dominant media culture. Time is spent less on intellectual advancement and scientific thinking and inquisitiveness and more on groupthink to connect technologically and stylistically within the culture.

COUNTERING THE MEDIA-EDUCATIONAL COMPLEX

The classroom is one of the only social organizations left in which sequence, social order, hierarchy, continuity, and deferred pleasure are important. This kind of educational structure should be maintained to foster the characteristics of students that help them to become successful in science. To do this, a curriculum should be centered around themes of writing, mathematics, and history, because the media works mostly against these areas. The media presents a fragmented curriculum, with information devalued. In the United States, the average person sees seven hundred thousand TV commercials, 260,000 billboards, and four hundred million TV programs by age 20.[5] Internet use also adds many more clips of information exposure. The result is a devaluation of information into tidbits of thought. This weakened flow of thought, in turn, weakens writing ability.

The mechanics of writing should also be emphasized to combat the media curriculum. Texting and fragmented TV clips actually emphasize shortened forms of writing to the point of interfering with smoothness in the writing process. Spell-check and instant thesauruses make writing seem easy but actually prevent the process of discovering one's own style by shunting its development into accepted conduits. In the past, writing developed by editing and reediting, with parents, colleagues, and teachers guiding the style and mechanics. Overuse of technology relies on the computer to simply guide writing according to certain algorithms and sequences (e.g., if a word is not in the database, then it is off-limits for the newly budding writer). Language should stress the understanding of metaphor in order to make the inferences needed in science learning.[6]

Schools should emphasize that knowledge is not a fixed entity, but rather a stage in human development. The uncertainty of knowing science is a characteristic of the scientist that makes him or her successful. It fosters a desire to go on questioning. Without this uncertain approach, knowledge is seen as static, uninteresting, and unquestioned. It is true that the Internet gives access to information—more and quickly—but are individuals alone able to determine whether that information is valid or should be questioned? The answer lies in fostering a curriculum based on critical thinking. The tools to judge scientific information, as shown earlier in this book, are not thoroughly incorporated in most science education programs. Superficially, all schools advocate such goals. However, the drive to present science with technology or with magic tricks and trinkets takes time away from the heart of science—the scientific method. The scientific process anchors science facts together to give coherence to science. Technology, when used correctly and under the guidance of instructors capable of judging science claims, can be an excellent resource for teaching science.

Schools should stress social cohesion and group work, particularly in laboratory settings, in light of the isolation technology brings to society. TV, Internet, and even radio require a focus on another course of communication that is not between people, but between the media and a solitary person. A person accesses the Web individually and not within a group. Thus, there is no chance for discourse or exchange of ideas. Science curriculum should mirror science in society—it takes place within a community. Ideas are shared and built upon in groups. Science classrooms should emulate this structure to foster critical thinking within a community. Fragmentation by the media-educational complex weakens science because of this isolatory mechanism. Without practicing collaboration and cohesion within groups, actual communities of scientists may be more difficult to develop. Chapter 9 discusses the role of the scientific community in establishing responsibilities for scientists and in the development of professional norms. Science curriculum should foster this kind of interaction.

Technology and the media can enliven and improve science teaching: Excel and statistical programs help streamline research and gain mathematical understanding rapidly from data; video clips and PowerPoint presentatives attract students to science lectures; and medical advances save lives. This chapter should not be seen as an opposition to technology or media in society. Instead, it should help readers understand *how technology uses us* as much as how we use technology. While media and technology savviness are important aspects of employment and social life, they cannot develop a person's weltanschauung, or philosophy of life.

DISSONANCE BETWEEN
TECHNOLOGY AND SOCIAL PSYCHOLOGY

At times some scientific advances are several stages beyond which society is ready. Discussion of the media-educational complex in this chapter exposes the structural threats of technology to science education and thinking. While technology is very much ingrained within our society, some advances may be blamed for serious problems, especially when people are unready for the advances.

Keyless ignition automobiles are becoming an option more frequently offered by car makers. The purpose is to make starting cars easy. Without a key there is less to think about and one less item to lose. However, three deaths—two in Florida and one in New York—indicate that the technology may be to blame for some people leaving their cars unattended in garages because they forgot to shut the engine off.[7] Accustomed to shutting a car down using a key for more than a century, society may have difficulty changing its reliance on the keyed ignition. Mary Rivera of New York, a college professor and former school superintendent, says that she suffers permanent brain damage as a result of the keyless ignition feature on her 2008 Lexus. Unattended, her car was left in her garage still running, causing carbon monoxide buildup in her house and killing her companion, Ernie Codelia. Rivera's lawyer contends that adequate warnings and checks do not come with the feature of keyless ignitions.[8] Toyota offered the following statement:

> Toyota's electronic key system fully complies with applicable federal motor vehicle standards and provides multiple layers of visual and auditory warnings to alert occupants that the vehicle is running when the driver exits with the key fob. Electronic key systems such as Toyota's are neither new nor unique within the automobile industry.[9]

According to www.edmunds.com, electronic key systems were optional in more than 150 vehicles in 2010. While there are several variations for the option, the National Traffic Safety Administration is considering tighter restrictions regulating keyless ignition systems.[10]

Social psychology plays a major role in media and educational learning. The keyless ignition may prove to be an advance that is in opposition to learned norms of vehicular operation. The public may reject some technology, despite marketing. Some will prevail. If the technology is deemed valuable, it will elicit enough change to increase its demand.

There were many people who thought TV would be "just a fad" and that society would grow bored of it. Is technology its own victim? Will Facebook be rejected or replaced? Will the media-educational complex outrun its own monetary needs, or will it grow in its domination of people's ideas? Will STEM (science, tchnology, engineering, and math) education change for better or for worse? These questions remain to be answered in the future. But one focus remains: how will science progress be helped or hampered by the media and its alliance to the technology and educational industries?

7

Pseudoscience

"What's your sign?" and "I'm stubborn because I'm a Taurus" are more interesting conversation starters than enantiomer positioning for most people. Knowledge of pseudoscience is ubiquitous; everyone is able to speak about his or her ideas on astrology or acupuncture without having a scientific background.

As discussed earlier, science exists in the shadow of the media-educational complex. Thus, in the absence of an effective science education system to develop public science literacy, people turn to the simple and attractive ideas of pseudoscience.

Pseudoscience is loosely defined as a body of knowledge and methods presented as scientific but not grounded in the scientific process. Pseudoscience lacks the tools scientists use to determine truth. There is no statistical analysis, control of variables, regulation, and collaborative study. Science is based on verifiable empirical evidence, but pseudoscience is based on emotion and belief.

In a society taken in by media superficiality and instant gratification, pseudoscience has been allowed to flourish. An astrology sign explaining someone's behavior is more direct than a detailed case study of the person explaining how a behavior may be based on genetics and his or her relationship to his or her brother. In contrast, events are explained simply when appealing to a pseudoscience.

Humans have an inclination to blame happenings, both in their own lives as well as in the larger world, on something or someone larger than themselves. William Shakespeare described this as the "admirable evasion of whore-master man, to lay his goatish disposition on the charge of the star."[1] Shakespeare is referring to the study of astrology, or making predictions based on star alignments. Many pseudoscientists make immense promises to their clients. UFOs, witchcraft, low-carb diets, gluten-free diets, acupuncture, and dowsing offer answers to questions.

U.S. polls by several media outlets in 2004 and 2008 indicate high support for most pseudoscience areas. The numbers are almost on par with belief in religion.

111

Some results of the national polls show that about a third of Americans believe in ESP (extrasensory perception, 50 percent), haunted houses (42 percent), ghosts (34 percent), and UFOs (34 percent); and about a quarter accept things like astrology (29 percent), séances (28 percent), reincarnation (25 percent), and witches (24 percent).

The goal of this part of the chapter is not to debunk or even debate every pseudoscience area. Instead, it is to address the psychology behind a public acceptance of nonscience as science and to discuss this as a threat to scientific progress.

ROOTS OF PSEUDOSCIENCE

Pseudoscience has historical origins that are linked to the true sciences. Before the development of the scientific method, observations and phenomena were interpreted, in part, on belief systems. As discussed in chapter 2, "Science Is Arguing," philosophies of science grew out of many centuries of study and changes in society that led to our modern science branches. Modern medicine grew out of ancient homeopathy and other philosophically based medicines of the past. Astronomy arose from astrological observations, hydrogeology (the study of groundwater) has roots in ancient dowsing, and chemistry arose from alchemy. This tie to science appeals to the public but the difference, again, is that scientific rigor is lacking.

IS PSEUDOSCIENCE LEGITIMATE?

While alternatives to conventional medicine are rooted in ancient cures, their claims do not hold up to scientific testing. While used for centuries in Eastern cultures, alternative medicines are classified as pseudoscience because they have not been subjected to the same level of rigorous experimental designs as used in the medical profession. Acupuncture, as shown in figure 7.1, is the practice of inserting needles to stimulate nerves in the body and treat health problems. Modern science explains how it may work: smaller diameter nerve fibers transmit pain but large diameter fibers carry other nonpain information. Scientists believe that acupuncture may stimulate large diameter fibers, thus blocking the sensations of small nerve fibers and pain messages. However, this is only a plausible physiological explanation for how the process of acupuncture works. Data have not supported the explanation because it is difficult to measure large and small diameter fiber activity through EMG (electromyography) studies. In addition, in most studies acupuncture has shown no significant improvement in patients given true acupuncture versus those given "fake" treatment. However, in some studies significant pain relief was shown. One study, conducted by Joseph Helms, a physician with the American Academy of Acupuncture, was performed using acupuncture on forty women experiencing menstrual pain, giving some real treatments and some placebo treatments (needles placed in the wrong positions). The group receiving real treatments showed a 50 percent

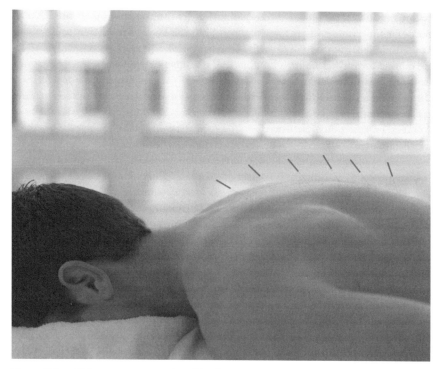

Figure 7.1. Old-new treatments for old pain: Acupuncture. Source: Creatas/Thinkstock.

decrease in pain and a decrease in the need for painkillers. However, the sample size of forty subjects is small, and a host of larger sample studies showed no difference in acupuncture versus placebo. To date, there is not significant evidence to support the claim that acupuncture helps any ailment.[2]

Alternative medicines, while favored by many patients for physical relief, advocate a philosophical rather than a scientific basis for health care. Acupuncturists, massage therapists, and chiropractors are licensed separately from medical professionals and medical practitioners do not work with and are not trained to coordinate with them. The holistic-based outlook for curing places these alternative practitioners at odds with the science-minded AMA (American Medical Association). Despite this, in the past decade chiropractors, massage therapists, and acupuncturists have gained increased coverage and access under health insurance and HMO policies.

The public continues to flock to these cures despite the lack of data in their support. There are two reasons for this. One is that many medical musculoskeletal problems improve on their own without intervention. Inflammation and cartilage are changing and moveable; thus pain is resolved without medical treatment. The second reason is that many conditions are psychogenic in origin. If the patient "believes" they are being treated, then they start to feel better. Migraine headaches and back pain often have roots in stress and anxiety. There are some solid studies supporting chiropractic and acupuncture work, but the body of research is much more limited than that of traditional medicine.

Another pseudoscience with roots anchored in our science history is astrology. Astrology uses the position of the moon, sun, and other stars at the time of one's birth to determine one's personality traits.[3] Astrology is so popular that I was offered my first college teaching job (I found out after the interview) because I was a Taurus. I was confused because I thought a Taurus's stubbornness would be a poor asset to a department. However, I was thankful and accepted the position, but I learned that getting work took more than just good grades, a research agenda, and a focus on teaching.

Two studies have substantively debunked this way of thinking: John McGervey determined the "sun signs" (the astrological period in which one is born, e.g., Taurus or Libra) of thousands of famous figures: politicians and scientists. He found no statistical tendency for members of either profession to be born under an expected sign more than any other sign.[4] A second study, by Shawn Carlson, conducted a controlled experiment in which famous astrologers were asked to correctly match the personality traits of subjects with their astrological sign. Results again showed no statistical accuracy in predicting a person's sign based on their personality.[5]

An ancient, related branch of astrology includes psychic prediction, which claims the ability to foretell future events. Our accepted understanding of physics makes the transfer of information from the future impossible.[6] However, the media-educational complex elevates the predictions of Nostradamus, the "great" psychic predictor of the sixteenth century. His poems allegedly foretelling future events are actually unclear and vaguely written. This strategy allows almost any event to have been predicted by his writings. He was a clever charlatan, foretelling world history events using certain verbiage that is widely open to interpretation. One of his poems supposedly foretelling the fall of Germany at the end of World War II follows:

> Animals ferocious with hunger will swim the rivers,
> The greater part of the armed camp will be against Hister;
> The great man will be carried in on an iron cage,
> When the German child watches the Rhine.[7]

While ostensibly alluding to the march into Germany by Russian forces in 1945 and "Hitler" being named, the verse must be more clearly studied within the historical setting of the writing. At the time of the poem's writing, the Turks were attacking from the southern Balkan region, but the French and Germans were suspicious of each other across the Rhine. Thus, the Germans were watching the Rhine River border with France, but the Turks were a threat on the southern border, the Danube River. "Hister" is the Roman name for the lower Danube, where the Turks threatened, and does not at all refer to Adolf Hitler.[8] Unfortunately, the media choose to present Nostradamus's poems as controversially plausible psychic prediction. Instead of appealing to a historical literacy or holding his predictions up to the scientific process, the media-educational complex chooses the popular and simple route of creating a mystique around Nostradamus. Nostradamus is commonly presented as a gifted fortune-teller, and few books or programs will expose the reality of this pseudoscience figure.

In modern times, a less socially acceptable form of pseudoscience than the others presented is dowsing. Dowsing is the practice of finding underground water for well digging by using a forked instrument or stick. Hydrogeology is the modern scientific study of groundwater. The importance of clean water availability and access to it cannot be overstated. Water's scarcity has limited populations since the ancients, and our society may be headed for a water scarcity crisis outside of the United States.

In dowsing, the dowser holds a stick, walking around until the stick moves, seemingly uncontrollably, over the site where water is to be found. Many wells have been found with the use of dowsers, and people swear by their efficacy. If water is found, an area can develop, so its importance is enormous. In the United States drilled water wells are very expensive ($4,000 to $20,000). In some areas of the world wells cannot be drilled or potable water cannot be found at any cost. Nonetheless, when dowsing is subjected to systematic scientific testing, it fails.[9]

A main reason for the success of dowsing in certain areas is that the underground water table is distributed relatively uniformly under the ground's surface in many areas. Evon Vogt and Ray Hyman give an example of a 100 percent success rate for dowsers in a part of Alabama. Expectedly, the groundwater is the same depth throughout that county, and the rate of success for nondowsed wells is also 100 percent.[10]

In general, water wells drilled without a dowser are equally as successful as those with one. L. Keith Ward found that of 1,823 doused and 1,758 nondowsed wells drilled in New South Wales, Australia, 14.7 percent of the dowsed wells were unsuccessful, while only 7.4 percent of nondowsed wells were. Surprisingly, dowsing had twice the failure rate of random drilling, meaning that there are better cues to water availability than dowsing.[11]

Alchemy is the study of converting one chemical into another. In ancient times, the main motive was to convert substances into valuable gold. While this is no longer a popular pseudoscience, it shows how society attempted to find easy answers to financial woes. If one could simply "make" gold, riches could be gotten quickly. The "lotto" culture of easily obtaining answers and solutions to problems is the attraction of pseudoscience. Alchemy at least led to the modern development of chemistry, with substances converted from one form into another. Oddly, substances can be converted to gold by destruction of the atom (nuclear fission), but it is certainly not profitable. The popularity of pseudoscience through the ages is based on certain sociopsychological underpinnings.

PSYCHOLOGICAL CAUSES AND IMPLICATIONS OF PSEUDOSCIENCE

Pseudoscience has a direct appeal to the public. It does not use the rigorous peer review process that establishes the truth and integrity of science research. It is a kind of scientific misconduct that thrives on the psychological weaknesses of the human condition. A certain amount of knowledge and reasoning in science, also termed *scientific literacy*, is lacking among the public, which allows pseudoscience to thrive.

People naturally seek a simple explanation for problems and questions. As discussed in chapter 2, "Science Is Arguing," a lack of science literacy requires that the public appeal to authority. Pseudoscience experts act as authorities to fill the void of science. The lower forms of reasoning, which are devoid of the elements of critically thinking about the topic, are sought. That is, the pseudoscience audience lacks the tools to evaluate scientific evidence to develop reasoned conclusions.

The importance of critical thinking and ways to improve it were discussed in chapter 5, "The Role of Critical Thinking." Without such abilities, pseudoscience draws on a groupthink mentality in which popularity wins arguments. This is the greatest threat of the media-educational complex. The media's goal is to obtain high ratings by appealing to the popular. In groupthink, the more people accept an idea put forth, the deeper the beliefs are ingrained into the culture. There is an old adage to this kind of propaganda: if you repeat a lie often enough, people will begin to believe it. Pseudoscience becomes stronger as more and more people believe it. In a media-driven curriculum, pseudoscience and myths flourish, obscuring valid science and reason. The greatest threat to pseudoscience is objective truth and scientific reasoning to support that truth.

Unfortunately, there is a great deal of money and power in masquerading as science. Both circus promoter P. T. Barnum and writer H. L. Mencken are said to have mentioned "the depressing observation that no one ever lost money by underestimating the intelligence of the American public."[12] The remark has worldwide application and is not particular to any culture. It is not, however, a lack of public intelligence, but a lack of scientific literacy (knowledge of science and reasoning) that victimizes the public. The fault lies in the mechanisms keeping the public from developing scientific skills.

Pseudoscience makes claims based on popular or even political demand. Throughout history, many governments have used propaganda to shape public opinion in support of various ideologies and political objectives. Radio broadcasts, posters, films, print publications, exhibits, and educational and cultural exchanges were all part of a broader program designed to manipulate public opinion.[13]

As seen in Einstein's poem earlier in this book, individuality and skeptical thought, as derived from critical-thinking strategies and tools used by scientists to analyze studies, drive science and creativity to unlimited bounds.[14] The push for individuality is a main thesis of this book. Many of the areas of pseudoscience may one day be shown as valid upon further scientific testing. Our minds should remain open to new information as long as it possesses the rigor of the scientific process. However, a public without science literacy will be lured by the popular and the superficial to be flummoxed by the myriad of pseudoscience available.

8

Debunking Science Myths

Separating Fact from Fluff

Belief in both pseudoscience and popular science myths are rooted in the same cause—diminished science literacy. Each pseudoscience is a larger paradigm of thought than the many myths that comprise it. For example, astrology has many myths associated with it just as science has many facts supporting its structure. There are misconceptions and myths in almost every area of thinking: history, philosophy, science, and plain old gossip, too! Science myths are so prevalent because we are exposed to scientific questions in everyday life. People have a natural curiosity for how things work and how life fits within the larger universe.

Some myth busting is easy in science because facts are very clear to present and verify. Is a baby bird rejected by its mother after a person picks it up and places it back in its nest? Simple observations of the behaviors of mother birds determine the answer. However, some myths are much more difficult to debunk because of their innate complexity. Is high-fructose corn syrup making America fat? This has an answer, but there are many facets to the question.

The purpose of this chapter is to verify or debunk some common ways of thinking about scientific phenomena. It will show the science behind whether a variety of common science myths are true or false. Most science myths fall into four categories: nutrition myths, nature myths, an ever-growing set of medical myths, and space and motion myths. The first is so popular because of the modern emphasis on body image.

This chapter should provide a useful foundation for science literacy. The overview reflects a changing emphasis established for science education by the National Science Education Standards set forth by the National Research Council in 1996 and the Benchmarks for Science Literacy by the American Association for the Advancement of Science in 1993. The focus on key concepts and the use of scientific arguments are both central themes of this chapter. Through the study of myths and the science that underlie each topic, a synthesis of the most important areas of science literacy, as described by the standards, will be treated: life, and physical, earth and space science content; inquiry and reasoning to answer scientific questions; properties and changes of properties in matter; motion and forces; transfer of energy; structure and function of living systems; life changes; genetics and evolution; behavior and diversity of populations and ecosystems; interdependence of organisms; structure of Earth's systems; Earth's history and the evolution of the universe and solar system; and technology and science processes.

A link to the national standards and a parallel with the content in this chapter helps to promote science literacy by answering some questions that many of us, including this author, require a bit more knowledge about. Debunking myths, and using the reasoning and tools scientists use to do so, creates an overview of science principles that are necessary for understanding nature's realm.

NUTRITION MYTHS

Every day in the news, headlines depicting an obesity epidemic in the United States are seen: "Chubbing up of America!" "Obesity is a nation's crisis," "More Americans choose doughnuts." The many health issues associated with the changing demographic of weight are real problems in our culture. Consider that in 1912, only 5 percent of people were overweight or obese. In 2012, over 65 percent of Americans fall into that category. Natural questions are "Why is this happening?" and "What can I do to prevent it in myself and my family and friends?" There is a genetic component to obesity, but there is a much higher environmental aspect. If being overweight were purely genetic, then the increases in obesity over the past one hundred years would not have been so dramatic; it would have been a problem long ago as well. People one hundred years ago were in the same gene pool of shuffled DNA as compared with today. Yet the obesity epidemic is soaring.

Obesity is a public health hazard, as many chronic conditions are related to it: type 2 diabetes, heart disease, high blood pressure, and arthritis. For every pound of fat a person gains, one thousand feet of extra blood vessels are made by the body to supply that fat. More vessels equal more resistance to blood flow. Thus, the heart must work harder to overcome that resistance with a higher pressure exerted in the blood vessels. This increase in blood pressure is the physical explanation for why obesity is linked to high blood pressure. Of course, the situation is not hopeless—if a pound of fat is lost, the blood vessels are reabsorbed and blood pressure should decrease.

Of particular concern is the rising obesity rate among children. The Centers for Disease Control and Prevention reported that 12.4 percent of children ages two to five are obese, 17 percent of those ages six to eleven, and 17.6 percent of adolescents ages twelve to nineteen. Unfortunately, 30 percent of children who are obese will remain so as adults. While there are many causes of the increased obesity seen in our population, it is not unique to the United States. Rising obesity rates are occurring throughout the world.[1]

Obesity may be due to a number of factors. As mentioned, genetics plays a role; comparisons of monozygotic twins indicate that obesity is about 30 to 50 percent dependent upon the genes or heritable. In modern Western society, the lack of physical activity is a result of changing work lives. One hundred years ago, more than half of the population was involved in agrarian work, which lacked mechanization and required large physical effort. Other jobs such as active construction work, trades, and manual labor (which is now often replaced by machinery) burned many more calories than the technology and office jobs dominating the workforce today.

Changing nutrition, however, remains a primary factor in contributing to increasing obesity rates in modern society. Eating foods that are high in sugar, energy, and calories clearly relates to increased weight gain trends. The availability of food (meats, cheeses, and processed products) in this era is unmatched. Even one-third of homeless people in the United States are obese. Access to food products of all types has resulted in a greater choice of foods for the public. Thus, many nutrition myths exist about the health benefits of these foods.

- High-fructose corn syrup is the cause of the obesity epidemic. TRUE (but not entirely)

This is one of the most complex myths to answer, and its physiologic underpinnings are not yet fully understood, but consider the facts. High-fructose corn syrup (HFCS) is made by first converting the starch in corn and wheat plants into glucose and then into fructose. This process is called enzymatic isomerization, and it results in concentrated forms of inexpensive, sweet, fructose syrup. It has replaced sucrose (table sugar) in most foods, sodas, and drinks in the past few decades. Fructose is in fact sweeter than sucrose and thus tastier. Fructose itself is a natural fruit-and-vegetable sugar, but the process of concentrating it into an HFCS product is not natural.

HFCS is metabolized differently than sucrose and glucose. Fructose is absorbed farther down along the small intestine and does not require insulin to be taken into cells. Other sugars require the insulin hormone to attach to a receptor on a cell to be absorbed, fructose enters the cell by a transport protein other than insulin. Thus, when HFCS enters the body, it does not stimulate the production of insulin and its related leptin hormone. Both insulin and leptin suppress appetite; without these hormones, it is possible that HFCS is taken into the body but a person remains hungry. This mechanism would be an obvious contributor to weight gain. In addition, brain cells do not have fructose's transport protein, and unlike glucose and sucrose, fructose cannot stimulate satiety signals to stop eating.[2]

One study from the University of California, Davis, compared how consuming beverages sweetened with 100 percent fructose versus 100 percent glucose affected hormone levels within twenty-four hours. The findings indicated that consuming fructose-sweetened beverages with meals caused a decrease in insulin and leptin hormones. Their findings also showed that fructose is more prone than glucose to becoming metabolized into fat by the liver. Thus, HFCS is associated with higher

LDLs (low-density lipoproteins), which are the bad fats in the bloodstream. This could lead to increased risk of developing cardiovascular disease.[3] This may be a myth if the mechanism described is not more clearly demonstrated through laboratory research. These are only plausible physiologic processes for HFCS derived from animal models. Further testing should be done to establish their truth.

I answered "TRUE (but not entirely)" to the question because of the contribution HFCS makes to obesity. People consume large amounts of soft drinks containing HFCS, with studies showing that children ages six to eleven drank about twice as much soda in 1998 as in 1977, while milk consumption over the same period declined by about 30 percent. One-fourth of adolescents reported an intake of sugared soft drinks of at least twenty-six ounces, which equilibrates into four hundred extra calories per day. For each sweetened soft drink consumed per day, the risk of obesity increases by 60 percent.[4]

Is there a motive in doing research to point the finger at HFCS as a cause of the nation's obesity? Who would have such a motive? Perhaps companies producing sugar and sweeteners that compete with HFCS are stimulating an outcry against their competition. For example, if naturally sweetened beet juice companies are put out of business by HFCS, then there is an intense motive to discredit fructose. One must consider the motive behind the research as well as the research itself.

The omnipresence of HFCS in our diets, however, makes it a dubious suspect in increasing our weight. Americans consume sixty-two pounds of HFCS each year. Bray et al. report that HFCS is the sole sweetener in sugared soft drinks and represents more than 40 percent of caloric sweeteners contained within other foods and beverages in the United States.[5] However, the answer to our original question must remain yes, maybe because there are more variables than constants to the issue. Decreased physical activity, increased access to all foods, and the fact that obesity rates are rising throughout the world (even in nations that do not use HFCS in their foods) indicate that there is more research to be done to answer a solid "TRUE."

- Blackened (or seared) meats cause cancer. TRUE

Upon searing, any grilled meat (poultry, fish, or beef) may develop chemicals that are carcinogenic. At high temperatures, natural chemicals in meat are converted to heterocyclic amines (HCAs), which are known to cause a variety of cancers. In addition, when the juices and fat fall onto charcoals, smoke is formed that may contain polycyclic aromatic hydrocarbons (PAHs). PAHs are known carcinogens and deposit on the meat as it is smoked. Few experimental tests were done on humans, but animal studies show the carcinogenicity of these troublesome compounds. Thus, smoked meats are also not recommended, as the smoke and PAHs concentrate on the outside of the meat. If you are going to eat smoked meats, try to remove the skin, which is likely to contain the majority of the carcinogenic chemicals.

• Drinking soda is linked to osteoporosis. TRUE

Soda replaces other more nourishing beverages, such as milk or orange juice that contain calcium to build strong bones and teeth. Osteoporosis is a thinning of the bone; it is due to a higher amount of osteoclast (bone-destroying cells) activity and a lower amount of osteoblast (bone-building cells) activity. It is also believed that the phosphorous in many sodas causes a leaching effect of calcium from the bones. Phosphorous is an essential component of bones in the form of calcium phosphate salts and hydroxyapatite $(Ca_3(PO_4)_2 * (OH)_2)$. As can be observed from the formula, both calcium and phosphorous are needed in these bone-building materials. However, when phosphorous is added to our diets in large amounts via soda consumption, it combines with calcium in the blood before becoming incorporated into the bone mass. Thus, calcium is not able to add strength to the bone because of the excess phosphorous. There is much debate on whether this phosphorous-leaching physiology is true. For example, if chicken contains a high amount of phosphorous, then why is it not related to osteoporosis?

I would answer "maybe" to the question if I had to definitively explain why there is a link. The mechanism is still not well understood. However, it is clear that replacing better beverage choices with soda is linked to osteoporosis as well as other ailments. The research establishing this link is well documented. For example, researchers at Tuft's University studied a large number of both men and women, finding that women who drank three or more phosphorous-containing sodas per week had a 4 percent lower bone mass density than women who drank non-phosphorous-based soda or less soda. Regardless of the mechanism, soda is linked to osteoporosis.

• Eating a separate macromolecule diet is good for weight loss. FALSE

Fit for Life authors Harvey and Marilyn Diamond contend that eating separate macromolecule meals (proteins, lipids, or nucleic acids) is recommended because the natural combinations poison the body and lead to weight gain and other ailments. This diet was well received by the public.[6] However, it is unnatural, and further research clearly shows that such diets would lead, over time, to nutritional deficiencies and health problems.[7]

There are many other diets that claim weight loss and nutritional benefits. A more recently popularized diet fad has been the Atkins diet. Robert Atkins emphasizes eating fat and protein and minimizing carbohydrate intake to lose weight and improve diabetes. Diabetes is indeed linked to carbohydrate intake, as these substances are quickly turned into sugars. The data do show initial signs of weight loss and a benefit to diabetics and prediabetics because the dieter is watching what he or she is eating and has a greater awareness about his or her food intake. However, actual weight gain occurs on a long-term Atkins diet. The initial benefits are probably because individuals are limiting their calories by not eating sweets, which are high in energy. In the subjects studied, increased blood lipid profile values (higher bad cholesterol levels) and higher blood pressure were also linked to following the Atkins diet longer term. Gluten-free diets are the latest fad, and while it benefits those patients with celiac disease, its link to weight gain, GI ailments, and allergies remains to be seen.

The golden rule of thumb in nutrition is that eating fewer calories than you require metabilically will result in weight loss. The Atkins diet is interesting metabolically: while a person eats a low-carbohydrate diet, the body will transform fats and proteins into the needed carbohydrates essential to life functions through anabolic reaction; these occur naturally in the body. The human body will simply make more carbs from fat and protein. Individuals beginning any new diet are always advised to consult with a medical doctor first. The link between arteriosclerosis and heart disease and several diets make them dangerous to an unsuspecting public.[8]

- Vegetables cause gas. TRUE . . . Why?

Vegetables, especially beans, have portions that are undigested in the small intestine and move into the large intestine for breakdown by bacteria. Hydrogen (H_2), carbon dioxide (CO_2), and methane (CH_4) gases are made in the large intestine, causing pain and abdominal muscle contractions to move the gas bubbles. The pain is almost always relieved rather quickly. Raffinose oligosaccharides are the likely culprit in a gassy situation. Beans and other vegetables contain this substance, which bacteria break down into large amounts of gas. The environment in our large intestines has a very similar atmosphere to early Earth conditions: methane, carbon dioxide, hydrogen, and sulfur gas (smelly); note that no oxygen is present, as in early Earth's atmosphere. Bacteria that first evolved resembled life in our intestines, developing from similar conditions.

- Olive oil is always good for cooking. FALSE (sometimes)

Yes, olive oil contains one of the best proportions of monounsaturated fats, 74 percent, which are related to good blood vessel and heart health and low proportions of saturated fats, only 14 percent, which is related to arteriosclerosis (hardening of the arteries). Monounsaturated fats contain a double bond and have less hydrogen saturating it than does saturated fat. In figure 8.1, a molecule of saturated fat shows that the carbon chain of a saturated fat is "saturated" with as many hydrogen atoms as is possible, while the monounsaturated fat has a double bond to prevent this saturation effect.

Saturated fat is often eventually deposited on the walls of arteries throughout the body, including coronary (heart) arteries, when in excess. This causes a change in the diameter of blood vessels with fat plaques. As shown in figure 8.2, in areas with fat deposits, the diameter of the vessel changes.

When a liquid (blood) moves through a smaller width, it changes its speed, creating turbulence. Turbulent blood induces clots to form, which block vessels and result in strokes (in the brain) or heart attacks (in coronary arteries).

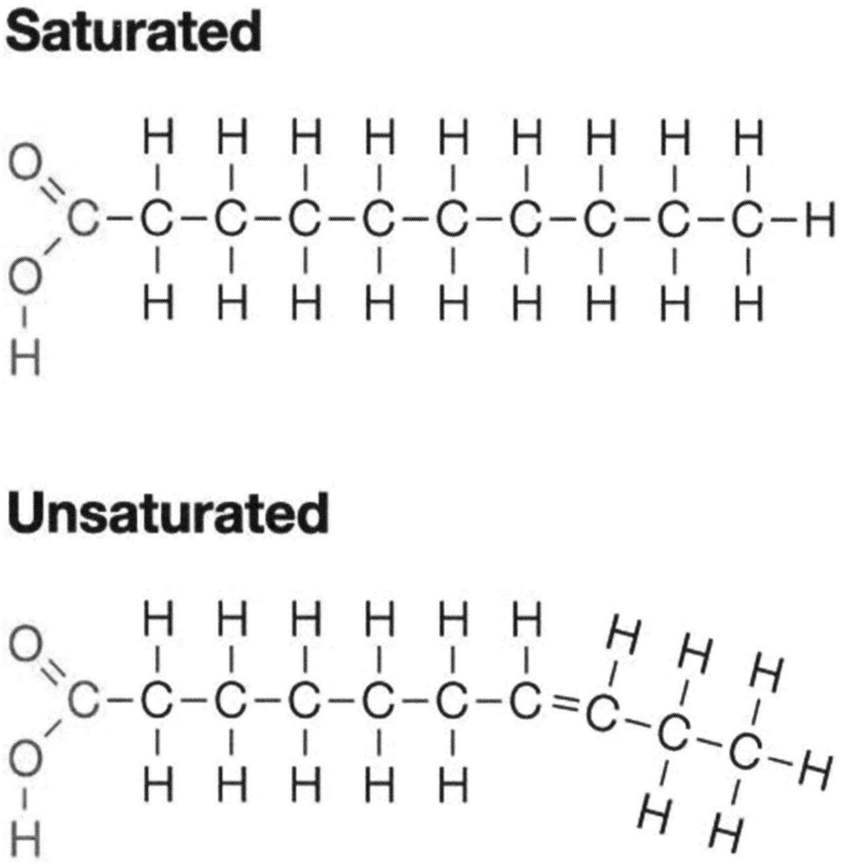

Saturated

Unsaturated

Figure 8.1.   Saturated fat versus monounsaturated fat.

Cooking with olive oil is great because it tends to increase the good cholesterol in the blood, high-density lipoproteins (HDLs), which are supposed to remove these fat deposits and lower the bad cholesterol in the blood, low-density lipoproteins (LDLs), which are directly related to artery fat buildup.

The answer to our question is sometimes because upon high heating of the olive oil (or any cooking oil), double bonds break, saturating the oil with as much hydrogen as possible, resulting in a saturated fat. The result reverses the effects of olive oil. A solution is to set the oil on a low heat while cooking and add a little water to temper the effects of the heat. Also, using grapeseed oil to cook may be better because it has a higher specific heat. Therefore, its double bonds break at a higher temperature than olive oils', and it works just as well in cooking for taste.

Grapeseed oil also contains a high amount of antioxidants, which are heart-disease and cancer-fighting agents. Antioxidants, as indicated by the name, prevent oxida-

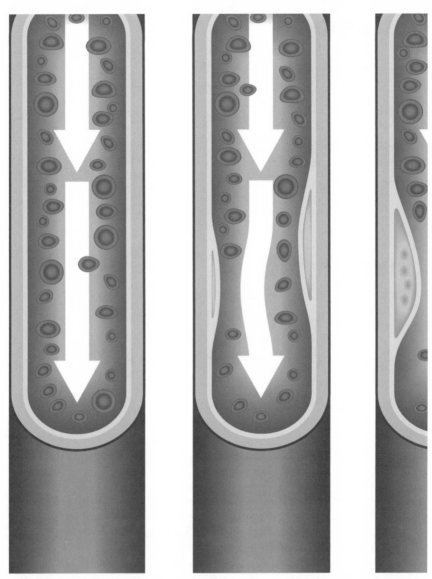

Figure 8.2. Turbulent blood flow in a clogged artery. Source: iStockPhoto/Thinkstock.

tion of foods. Oxidation causes a loss of electrons, which leads to free radicals. The hypothesis goes as follows: free radicals have extra electrons on them that are very reactive and cause damage to parts of the body, including DNA (deoxynboncleic acid) and vessel walls. If DNA is damaged, it leads to many problems, including genetic defects and cancer. If walls are damaged, it leads to inflammation of the walls and possible clotting, as described in the mechanism above. Clearly, eating a diet high in antioxidants would help prevent free radical formation. Foods high in antioxidants are garlic, blueberries, and raspberries; at the next level are onions and broccoli and then carrots and leafy greens. All of these high antioxidant foods are fruits and vegetables, which help fight disease. Inclusion of vegetables and fruits in our diet is an ever-present theme in nutrition.

- You should wait to swim after eating a meal. FALSE

There is no evidence that muscle cramps occur during swimming as a result of food's digestive processes. This myth was perpetuated because there is some physiological truth behind it. The parasympathetic nervous system is a set of nerves that become activated with digestive processes. These nerves send messages to bring blood and energy from other body parts toward the gastrointestinal tract and associated organs. Energy is taken away from muscles and heart function to go toward digestion. However, there is no evidence to support muscle dysfunction or heart problems during exercise simultaneous with digestion.

- Skinny people are healthier than overweight people. FALSE

Generally, medical experts suggest that a person's BMI (body mass index) is the best measure of a person's weight because it includes height and makes a prediction on a sliding scale as to one's health risks based on weight. BMI measures have come under fire for oversimplifying a person's weight as an expression of their health. Critics are correct; it is the abdominal and organ adipose (fat) tissue that is most dangerous for cardiovascular and even cancer risks. There are generally two body shapes: pear and apple. A pear gains weight around the hips and thighs whereas an apple gains weight around the midsection. Apples tend to have more abdominal adiposity than pears, and the apple shape is more strongly correlated with cardiovascular and cancer risks. A better measure of one's weight health is a simple measure of waist circumference. If it is a smaller number than half of one's height, then that is a good reading; the smaller the better.

- Putting hot food in the refrigerator ruins the food. FALSE

There is no evidence that allowing food to "stew in its juices" at room temperature adds flavor. The same taste will evolve even in the refrigerator. Hot leftovers should be put immediately into the refrigerator to prevent bacteria and other microorganisms from growing. During the cool-down phase to room temperature, food moves through many temperature ranges, from 57°C to 15°C (135°F to 41°F), also known as the "danger zone." Many types of bacteria are able to grow at these temperatures, and cooling foods provide nutrition and a good home for these critters. *Salmonella*, *Pseudomonas*, and *E. coli* grow well at 37°C, and *Staphylococcus* (often found on skin surfaces) grows at 25°C. Many bacteria are able to grow at refrigerator temperatures. However, the refrigerator slows *Pseudomonas* considerably, so it is of little concern for weeks. I see only one frugal flaw in placing hot food in the fridge: it will take more electricity to bring the meal down from a higher temperature than from room temperature.

- Vitamin C prevents colds. FALSE

Many studies have conflicted in their results on the effect of vitamin C on common cold prevention. The common cold is due to the rhinovirus, which invades the body through mucous membranes in the upper respiratory tract. The virus is delicate and can only be transmitted in this manner. The Mayo Clinic performed several analyses showing no difference in cold prevention or duration between those taking vitamin C and those who did not. Also, there is no physiological method yet discovered by which vitamin C could work to boost immunity or attack the rhinovirus. In various reviews of research spanning more than eleven thousand subjects, those taking 200 mg of vitamin C per day had as many colds as those who took no supplements.

This myth originated in the 1930s in Linus Pauling's book *Vitamin C and the Common Cold.* Pauling reported his belief and evidence that vitamin C prevented colds, and he encouraged people to take 1,000 mg a day to ward them off. Most medical experts dismiss this advice. However, foods rich in vitamin C also contain other nutrients and antioxidants that are healthy. I recommend vitamin C in the diet because it comes along with many other healthy vitamins and nutrients. It is better to eat an orange to get vitamin C than to eat potato chips. A large glass of orange juice contains nearly 100 mg of vitamin C, and it is certainly better than soda.

- The Food Pyramid and the Food Plate are good guides to nutrition. FALSE

Our species, *Homo sapiens,* has a metabolism (totality of chemical reactions occurring in the body) that has developed as a result of over 180 million years of mammalian evolution, sixty-five million years of primate evolution, five million years of hominid evolution, two million years of our genus *Homo,* and about 150,000 years as *Homo sapiens.*[9] The human body is adapted to a lifestyle lost long ago, with a caveman/cavewoman diet very different from modern times. The prehistoric nutrition system is termed the Neolithic diet, and it includes foods that were eaten by prehistoric humans. Our genes are adapted to this kind of diet, and modern humans should follow this diet to be most healthful. Years of evolution cannot be ignored when understanding the obesity and diabetes epidemic in Western society. The Food Pyramid and Food Plate guidelines collide and conflict with the Neolithic diet that our evolutionarily metabolism craves and requires.

The majority of human prehistoric diets emerged from our nomadic lifestyle, through which mostly whole grains, fruits, nuts, and vegetables prevailed. Few opportunities to obtain large kills of meat and fat existed because of the difficulty in obtaining large game. Obviously, proteins and fats were a part of the Neolithic diet, but it scarcely comprised much of an average daily intake of the total calories.

Wild, small-game animals and insects were more prevalent and made up most of the prehistoric proteins.

A reconstructed Neolithic diet is at odds with both the Food Pyramid and Food Plate. The Neolithic diet contained large amounts of fresh fruits and vegetables; whole grain (unrefined) starches such as rice, tubers, acorn, and grasses; small-game meats such as frogs, birds, snakes, fish, and even insects; extremely scarce (and coveted) refined sugars in the form of honey; and few fats (in the form of small-game catches). In contrast, the current models for nutrition put forth by the U.S. Department of Agriculture (USDA) call for too much starch and fat and not enough protein, dairy, fruits, whole grains, and unrefined starches in comparison with Neolithic diets. This is disturbing because state and national food programs derive from USDA recommendations. More concerning, however, is the actual diet of most Western societies that, of course, are even further from the Neolithic diet, with fats, meats, and refined sugars dominating caloric intake.[10]

Preagricultural peoples led more active lifestyles than those in modern Western society, with caloric needs of about three thousand calories per day. Nomadic lives included hunting game and gathering vegetables, as well as caring for their young and building shelters. However, they consumed far fewer calories from fat, with meat consumed from wild game containing 4 percent fat, while present-day corn-fed livestock contains 29 percent fat. Our ancestors ate half of the fat we do but three times the protein.[11] Protein was obtained from less fatty natural game and from nuts and beans, while our protein is obtained primarily in higher-fat meats and dairy.

In fact, when was the last time you ate pizza? Most people will answer within a month's time frame. Consider the components of pizza, which contain drizzled cheese and oils. A fourteen-inch slice of pepperoni pizza contains 298 calories, with 37 percent fat, 47 percent carbohydrates, and only 14 percent protein. Compare this with a serving of deer meat, which contains only 32 calories per ounce and has 18 percent fat, 0 percent carbohydrates, and 82 percent protein.[12] This mismatch between modern-day food choices and our physiologic origins may be contributing to the obesity epidemic seen today in modern Western society.

The advent of agriculture contributed to the change in diet. With farming, the general health declined, mainly because the variety of food types decreased and only a narrow range of cultivated crops were available to the public. Instead of the many wild plants available during Neolithic eras, agricultural populations had 90 percent of their diet based on their crops. Roots, beans, nuts, tubers, and small game of the Neolithic era had far more vitamins and minerals than cultivated starches of crops.[13]

Of course, the food supply increased the number of individuals capable of being fed, so a population explosion was spurred by the development of agriculture. Agribusiness improvements make the carrying capacity (number of individuals an area can support) much larger than expected without technological advances in farming. In preagricultural eras, the population of the Earth doubled every fifteen thousand years. After the rise of agriculture, the population of the Earth doubled every thousand years or so, and it currently doubles every thirty-five. In 2000, the world's population hit 6.15 billion. It is now over seven billion and is estimated to be over nine billion by 2050.[14] The Plains states support the majority of U.S. food exports and provide nutrition to much of the world. Without technology-based farming, most of the world would starve. But is there a cost associated with our divergence from our evolutionary past?

- Eating salt causes high blood pressure. TRUE

Salt is a necessary part of the human diet. In marine aquatic environments, the opposite is true, and there is too much salt for living creatures. This requires specific excretion systems in some marine inhabitants to pump out the excess salt. Historically, as evolution led creatures from the sea to land, no means of obtaining salt were developed except through diet. Thus, salt must be obtained from the food we eat. Salt was a valuable traded commodity among humans; the ancient world viewed salt as more valuable than gold. Wars have been fought over it, empires were built upon it, and children were sold into slavery, all over salt. This is how the saying "not worth his salt" originated from Petronius in the *Satyricon*. Roman soldiers were paid salt rations, and the Bible has many references to salt as a commodity. The National Research Council of the National Academy of Sciences recommends that 1.1 to 3.3 grams of sodium (salt) be consumed each day. For every one thousand kilocalories of food in a person's diet, one gram of sodium should be taken in. A healthy salt intake should not exceed 4.8 grams per day. A few studies suggest that there is no relationship between salt intake and high blood pressure. This has perpetuated the myth that salt does not cause high blood pressure.[15]

Through an understanding of the laws of chemical and physical movement, this myth can be dispelled. In medicine, treatments for high blood pressure and recommendations for low-salt diets derive from an understanding of these laws. All matter, even in solid state, is in constant motion (at the atomic level) due to an intrinsic amount of kinetic, or moving, energy. Objects in motion close together bounce off one another. Kinetic energy keeps objects moving so that they spread out farther and farther. Chemists call this process *diffusion*. Particles and concentrations in liquids and gases tend to move from areas of higher concentrations (amounts) to lower concentrations until an equilibrium, or even spread, is reached.

The blood is composed of mostly water and dissolved salts. Salt changes the flow of water into or out of the vessels. If a diet is high in salt, more salt will make its way into a person's bloodstream. Water flows as the laws of diffusion dictate. As a rule, water follows salt. Given the high salt intake, water follows salt into blood vessels and increases pressure in them. Diffusion will move water from higher concentration areas (in body cells) to where it is lower (in the vessels) in the presence of more salt.

When water enters blood vessels, the higher blood pressure may cause damage, and this is the risk of sodium-induced hypertension. This relationship is responsible for many health ailments, including strokes, heart attacks, and kidney failure. Two-thirds of Americans consume more than the recommended intake of salt. Clearly, a relationship does exist between salt intake and high blood pressure.

Medications are difficult to work with because each person responds differently to drugs. There are three main types of blood pressure medicines: 1) vasodilators, which expand the vessels to lessen pressure; 2) beta blockers, which limit the excitement a person feels by reducing the effect of adrenaline hormone; and 3) diuretics, which reduce the amount of water in the vessels by increasing urination. While the human body does respond to the chemical and physical laws of science in the treatment of blood pressure, a patient's response to medicines, including blood pressure medications, is not always predictable or fully understood.

This demonstrates how error is inherent in science. Frequently evident in medical research, nonconformities can be particularly frustrating for both patients and doctors. However, lifestyle, diet, exercise, and health education may stave off hypertension in many people.

- Diabetes is a new disease. FALSE

Diabetes is not a new disease and, in fact, is a useful and natural sugar-sparing technique used in prehistoric times to conserve energy in times of starvation. Genes linked to diabetes are termed "thrifty" genes because they keep sugar levels high in the blood. Diabetes is a result of "thrifty" genes evolved over millions of years. Diabetes is defined as a disease characterized by a higher-than-normal level of glucose (sugar) in the blood (80 to 120 mg of glucose/100 mL blood). Under normal conditions, an animal consume sugar, activating beta cells of the pancreas, which makes the insulin hormone. Insulin stimulates the uptake of glucose into body cells and also causes the liver to store glucose as glycogen. When blood sugar levels drop too low, alpha cells of the pancreas secrete the hormone glucagon, which stimulates the liver to release sugar from the stored glycogen form. This negative feedback mechanism consistently maintains homeostasis of sugar to around 90 mg/100 ml blood throughout a lifetime.

Diets high in sugar and simple refined carbohydrates are linked to the development of diabetes because excess sugars "wear out" insulin receptors. Intake of food with refined sugars, such as doughnuts and cakes, elicits a surge in insulin and a docking with cells in the body. This wearing-down process is known as *down regulation* in endocrinology. Each cell in the body has proteins on membranes on which insulin attaches. After docking, insulin causes the target cell to take up glucose. Diets high in sugars wear out insulin receptors and cause insulin resistance. Diabetes type 2 (also known as adult-onset diabetes) works in this way; not to be confused with type 1 diabetes, which is an autoimmune attack on pancreatic cells that produce insulin. Both result in hyperglycemia (high sugar levels) but have very different mechanisms.

Damage to tissues and organs occur whenever glucose levels are too high or too low. In type 2 diabetes, hyperglycemia results in diabetes, with insulin unable to allow sugars to be used by cells. This keeps blood sugar levels high, and cells do not obtain needed energy; thus, they are starving. In the face of this situation, the body compensates by breaking down fats and proteins to halt the starvation process, releasing ketones. Ketone breath is fruity and denotes starvation and/or diabetes in patients. The danger of ketones is that they are acidic and lower the pH of blood, sometimes resulting in unconsciousness (diabetic coma) and even death. Clearly, these diabetic bouts damage nerves, blood vessels, heart muscles, and other organs. Alternatively, at low levels, for example, hypoglycemia takes place, with individuals suffering from weakness, disorientation, and even unconsciousness as the brain is deprived of needed sugars. This would be a disadvantage for prehistoric individuals who missed meals when unable to easily obtain food in harsh environments.

A mutation to prevent the conversion of glucose to glycogen would be beneficial for maintaining sugar levels during starvation conditions. These "thrifty" genes spare the sugars in the blood, keeping it available for later use. Consider the Neolithic diet, in which calories may be hard to find at certain times of the year; for example, when there is little game to be found. The individual with "thrifty" genes would benefit because normal circulating levels of sugar would be maintained longer to retain life functions for hunting and gathering. James V. Neel, a geneticist at the University of Michigan, discovered the "thrifty" gene sequence in some human populations. When faced with starvation conditions, natural selection would have favored such gene sequences. The Pima Indians of Arizona, he found, were not only resistant to insulin's effects of taking up sugar from the bloodstream but also retained more fat storage. This would help the Pima endure longer periods of reduced food availability and starvation conditions.[16]

The advantages of having "thrifty" genes to store more fat and keep circulating the blood sugar available were important to survival for populations during preagriculture times. However, in modern society, with the availability of food much increased and a change from hunter-gatherer lifestyles requiring more energy, people with "thrifty" genes are more prone to obesity and diabetes. In fact, about half the Pima Indians have diabetes, and about 95 percent of those individuals are obese.[17] Are diabetes and obesity on the rise due to the dissonance between evolved metabolism and modern diets? Can lifestyle changes improve these maladies? Studies indicate that a return to more traditional lifestyles and diets could improve society. Pimas practicing traditional lifestyles in isolated parts of the Sierra Madre mountains of Mexico have significantly lower rates of diabetes (8 percent) and obesity (rare) as compared to those of the modern U.S. Pima Indian population.[18] This may be a case study to guide changes in our approach to combat obesity and diabetes. A diet rich in variety and whole grains, fresh vegetables and fruit, and low-fat protein sources has abundant support in science as a recommended diet.

- Coconut oil is good for your health. FALSE

Coconut oil has an unusually high amount of saturated fat and minimal amounts of monounsaturated fat. It contains about 92 percent saturated fat, 6 percent monounsaturated fat, and 2 percent polyunsaturated fat. Recall that saturated fats contribute to increasing bad cholesterol (LDLs) in the body. Compare coconut oil with olive oil, which contains only 14 percent saturated fat and 74 percent monounsaturated fat with 11 percent polyunsaturated fat. Olive oil is much better for heart health because of its high amount of monounsaturated fat, which is linked to good HDL cholesterol. While coconut oil's saturated fats are medium-chain triglycerides (MCTs), which are linked to improving weight loss, these studies are only found in animal models. Human studies do not support the benefit of MCTs over other saturated fat in risks for heart disease and obesity.[19] Long-term studies on humans do not exist to support the claim that coconut oil is any different in its effects as compared with any other saturated fat product.

Research on Pacific Island and Asian populations, with diets naturally high in coconut oil, indicates low rates of cardiovascular disease and cancers. However, these populations have a more active lifestyle, eat primarily vegetarian foods, and have little access to fatty meats. Making a comparison to island people's diets is unrealistic because there is little control of the variables between the groups. Is there a movement afoot to portray coconuts as healthy? Is the media doing so for the coconut industry or to bolster foreign trade with tropical regions? The evidence is strong that the saturated fat in coconuts is bad for our health.

- Radiation in fish from the Japan earthquake is harmless. FALSE

The natural world has a constant background of radiation emitted from radionuclides (atoms with unstable nuclei) as they decay. This background level comes from a combination of constant bombardment of the earth by cosmic rays as well as radioisotopes of common elements, including those that make up the very backbone of life, carbon (carbon-14) and hydrogen (tritium). Radioactive elements such as uranium are natural parts of the planet and decay over long periods, giving off radiation. Radioactive isotopes, however, exist in small proportions to more stable isotopes; thus, the diluted natural radioactivity poses little danger to us.

Radionuclides can be useful for assessing the age of a substance due to known decay rates, also called their half-life. Carbon-14 is used to date organic remains from archaeological sites. Radioactivity due to nuclear weapons testing has been detected in soils and sediments since 1954, with peak concentrations of cesium-137 (a radioactive element produced in nuclear fallout and reactors) in sediments corresponding to the 1963–1964 nuclear tests performed en masse just before a test ban treaty went into effect. This radionuclide can be used to assess trends in historic atmospheric pollution rates.[20]

Radiation hazards result, however, when radioactive nuclides are concentrated outside a planned containment. Any amount of increased radioactivity is a risk for human health. Knowing this, the question really is "What level of radiation is healthy (tolerated by the human body), and what amount is too much?"

Fish first absorb radionuclides into muscle tissue. Scientists recently reported that Pacific bluefin tuna migrating from Japan to southern California waters contained more radioactivity than in the past. Every single fish tested showed increased levels of cesium-134 and cesium-137. In tests on fish from 2008, no radioactivity was seen in any fish from Japan. This rise in radioactivity is attributed to fallout from the nuclear disaster at the Fukushima Daiichi power plant in Japan in March 2011, following a massive earthquake and tsunami.

While sushi (often made from the Pacific bluefin tuna) is a favored food throughout the world, its risks are not given much publicity. Official statements claim that the amount of radioactivity in Pacific bluefish tuna are only one-tenth the level the United States and Japan consider harmful to human health. However, an increased level should be considered harmful because it is the amount of exposure that contributes to DNA and human tissue damage, increasing risks for cancer and genetic diseases.

If we then consider nuclear power, having led to this dilemma, to be unsafe, what is the alternative? More than half of our energy is sourced by coal power, which has its own safety concerns. Emissions of mercury, carbon dioxide, and sulfur dioxide are considered dangerous to humans and earth. Movement to alternate sources of electricity has been limited in the past thirty years. Only 1 percent of our power comes from wind and solar energy combined; nuclear power provides almost twenty times that amount of power. U.S. policy indicates that the United States is serious about maintaining potentially deadly sources of power—nuclear, gas, and coal—to drive our economy. There are many reasons for this, which are beyond the scope of this book, but the consequences are obvious. The search for clean and safe energy is a necessary goal for future science. Whether eating fish from Japan with radionuclides is healthy or not is only one question emerging from this tenuous topic.

NATURE MYTHS

Nature and the world around us are fascinating, and we are linked to it in so many ways, whether living in a city or the countryside. Our history emerges from the sea and salts, bringing our minds always closer to the environment to gain better understanding. It is with awe that we sometimes misinterpret nature and its cues. A selection of myths and truths about our nature exist, but the small subset that follows gives a focus and a guide to some common themes in nature.

- Evolution leads to the "strongest" organism surviving. FALSE

Evolution is the change in species over time; it is a product of the pressure by nature to select out the weak and keep the creatures best adapted for that particular environment. The strongest do not always survive. The dinosaurs were very strong, according to fossil prototypes, but they were selected out of the community; they did not best adapt at some point in the past. Evolution does favor the devices of life, which lead to more offspring; these are numerous and multivaried but always lead to greater numbers of children. This is termed *reproductive success*, which equals the number of live young that an organism produces.

Life's devices do not need to be complex. Another related myth is that evolution leads to increasing complexity. This is not really the case. The best-adapted creatures are often the simplest. Prokaryotes (bacteria) have been around for most of Earth's history, 3.5 out of the 4.1 billion years of the planet. Stromatolites are buildings of bacteria with layers of fossils from as far back as life's history. Prokaryotes remain very competitive because of their simplicity. They contain nearly nothing, only a primitive nucleus, a cell membrane, and some cytoplasm. There are no fancy organelles such as those plants and animals have. This limits the amount of things that can go wrong with the creatures.

Microbes cause many problems for humans; diseases were rampant throughout our history, and viruses and bacteria were often the cause. They divide at rapid rates, allowing them to mutate and evolve at a much faster pace than humans. They can divide in seconds, but it takes a generation (twenty-five years) for humans to have children and increase their variation. In these new offspring, newly created organisms have new variety. Variation is the key to any species' success because it allows some members to survive new conditions. For example, when some people were able to survive the plague of the Middle Ages, the human species lived on. If everyone were of the same genetic code, all individuals would have died, leading to extinction. Because microbes are able to divide quickly, they are better adapted to change than more complex creatures, such as us. Based on this, I predict that the microbes will be our doom!

Consider the old Volkswagen Beetle, which had no air conditioning, power locks, power windows, or power brakes. It ran and ran for decades without problems. The VW is much like a prokaryote. Alternatively, more expensive and complicated automobiles contain many features that can break down. Anyone who has had a check engine light turn on appreciates the aggravation in finding the small problem causing an emissions issue. The complexity of humans and other creatures are their downfall. Over 99 percent of all life that inhabited this planet is now extinct! Complex life does not necessarily survive better, and I would certainly predict that prokaryotes will outlive humans.

- Dinosaurs are extinct. FALSE

Dinosaurs are not extinct because they share very similar DNA with birds. Birds are direct descendants of dinosaurs. Feathers from dinosaur fossils were found in 1860, and the link continued with more than twenty species of dinosaurs. Studies of fossils from dinosaurs, namely the *Tyrannosaurus rex*, show that there is more relatedness between dinosaurs and birds than with reptiles, amphibians, or any other creature. Collagen fibers, which are ropes of proteins in the soft tissues of animals, were studied to compare compositions between the species. Obviously dinosaurs are no longer roaming the planet, but the greatest component of their DNA remains—the birds.

The extinction of the traditional dinosaur species we have come to know is a hotly debated topic. It is a debate between geology and biology as the cause of extinction. It has long been held that an asteroid hit the Earth about 65.5 million years ago, causing a major shift in climate. Dust from the impact led to less sunlight, fewer plants, and thus less food for dinosaurs and other species. Fossil evidence dating back to that period shows higher-than-normal amounts of certain materials, including iridium, indicating meteorlike hits. The layer of soil containing these particles is known as the K-T boundary (Cretaceous-Paleogene boundary), and it correlates with a high extinction rate for many species types. The large crater on the Yucatan Peninsula is thought to be the evidence of this meteor impact.

However, emerging theories implicate a viral or other parasitic infection as the cause of the dinosaur extinction. It is known as the hyperdisease theory of dinosaur extinction; a microbe rapidly evolved to kill off other living creatures during the time period. The power of epidemics to destroy cultures and species has historical grounding. Some historians claim that 70 percent of Native American Indians died due to the disease brought over by Europeans and not bullets, as is often depicted.

- There is a biochemical challenge to evolution. Intelligent Design is valid. Both FALSE (for now)

Evolutionary evidence is based on just that—evidence. There is no time machine to help us go backward and observe what really happened to life. Scientists reconstruct the past through evidence. Similarities in anatomy among many species (e.g., the whale fin, human hand, and bat wing have the same bone structure) show relatedness and imply a common ancestor. Fossil records show that species change over time. Tracing molecular proteins and genetic relationships among organisms shows strong evidence for evolution leading to current-day life-forms.

However, it would be remiss to merely discard alternate working hypotheses: Intelligent Design argues that the discovery of life's complexity means that it would be too difficult for life to have slowly evolved in response to the environment. This is a contentious proposition because such a claim is loaded with political and religious connotations. Michael Behe, a biochemist, in his book, *Darwin's Black Box: The Biochemical Challenge to Evolution*, argues that life's chemistry is too complex to have emerged because precursor steps were not beneficial.[21] He cites, for example, the clotting process, which has over two hundred steps. He claims that each step would not benefit an organism, but all of the steps together create a clot (which is the benefit evolution would favor). Behe also calls the eye, ear, and bacterial flagellum too complex to have evolved in one swoop; a term he uses is that they are *irreducibly complex*.

Most of the scientific community disagree with the hypothesis of irreducible complexity. There are potential benefits of intermediary steps in blood clotting, according to some scientists.[22] Precursor steps aide an individual's chance of survival so that those steps would have pressure to remain in the genes. In this way, more and more complex systems could be built upon. The steps to building the eye, ear, and cilia are shown by some scientists' writings to have benefits in organisms.

The preponderance of evidence supporting evolution at this point requires an answer of "FALSE" and places irreducible complexity in the realm of pseudoscience. Intelligent Design theory models itself along similar principles, so it must also be classified as such. However, if evidence for evolution were to be better refuted, then irreducible complexity would reasonably challenge evolutionary paradigms. For example, if the precursor stages of complex systems were shown not to benefit an organism, then Intelligent Design would gain favor. Reasoning is not about politics but about data and the support of hypotheses. The reason this myth is particularly sensitive is its underlying religious and social implications; it hits at our creation and our importance in the universe. It is an important myth to discuss because of its ultimate value in our philosophical well-being.

- Evolution explains the origin of life and the origin of the universe. FALSE

Evolution explains neither of these fascinating origins; in fact, it explains life only after it evolved. The Big Bang theory attempts to explain how the universe formed. Current models to explain the origin of life contend that life is able to form spontaneously from simple precursor models. Let's trace the development of matter from its early stages. About 13.75 billion years ago, the universe expanded from an extremely dense and hot state; it continues to expand in what is termed an *inflationary epoch*. The Big Bang theory has sufficient evidence for the start of matter because many of the observations of movement and particle attraction make sense based on the existence of such a past event. Matter continues to explode and expand, with our local sun a product of this Big Bang. Physicists calculate that over 70 percent of matter in the universe is dark matter. Dark matter is not able to be seen but is detected based on its gravitational pull. All objects exert a gravitational pull on other objects. There is too much gravitational pull on objects in the observed universe to account for it; thus, there must be some other forms of matter and energy.

It is estimated that dark matter makes up 23 percent of the universe and its affiliated dark energy makes up 72 percent. Atoms and matter detectable to humans, as we understand it, compose a smaller part of the universe. The organization of structures smaller than subatomic levels is given in figure 8.3.

Physicists are studying particles within the atom known as quarks. Quarks are parts of neutrons and protons (particles inside of atoms). Quarks are hypothesized to have even smaller parts known as *superstrings*. Superstrings, or strings for short, are one-dimensional loops of matter that vibrate as an infinitely thin rubber band. This view of matter is known as String Theory; it has been mathematically but not experimentally shown because it cannot be directly detected. Quarks and strings are still in the realm of science, but without more tangible evidence, the science is developing. All of this material emanated from the Big Bang.

The origin of life itself is also not explained by evolution. Origin-of-life scientists explain that the primordial mixture of chemicals on early Earth (about 3.7 to 4 billion years before present, or BP) could give rise to macromolecules. These chemicals have been shown, in the laboratory, to self-assemble and form primitive cells known as *coacervates*. These "cells" are separated from the external environment by a layer of phospholipids. Phospholipids have parts of their structure that are attracted to water

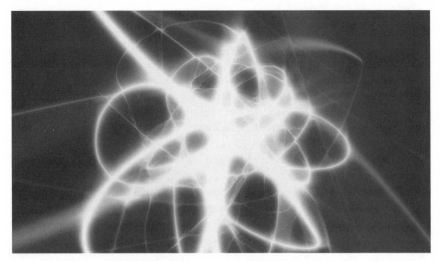

Figure 8.3.   Strings and quarks. Source: photos.com/ThinksStock.

and parts afraid of the water, or hydrophobic. Hydrophobic areas direct themselves away from water in ways reminiscent of "oil and water do not mix." Thus, a circle of phospholipids forms. The creation of life's precursors is well documented. It is often illustrated as a first experiment in college cell biology laboratory courses. The making of genetic material is a bit more elusive. For a cell to be living, it needs to be able to reproduce itself. It is thought that clay and negative charges aggregated genetic materials, which then eventually made their way into coacervates, creating a reproducing cell—life's first cells. Remember that all of this occurred over many hundreds of millions of years, so the chance of such events happening is quite plausible. Obviously it is difficult to imagine millions of years, just as our national debt is hard to wrap one's head around. It is necessary, though, to appreciate the time it took to yield evolutionary products: life.

- Sharks do not get cancer. FALSE

Shark skeletons are mainly composed of cartilage, which is a tissue made of roughly 80 percent water and has pliability. For the last one hundred years, scientists have studied cancerous tumors in sharks. The first shark tumor was recorded in 1908, and scientists have since discovered benign and cancerous tumors in eighteen of the 1,168 species of sharks studied.[23] Sharks do have antiangiogenesis factors, which prevent blood vessel growth in tumors and thus inhibit cancer cells from growing.

Many nutrition stores carry supplements with cartilage from sharks claiming to prevent or cure cancer. After much analysis, there is no evidence to support such claims. Cartilage does not have healing properties and is merely a normal part of humans, as in all animals. Cartilage is a cushioning material at the ends of our long bones (in the knees and hips, for example) and in between the vertebrae of our backbones to provide flexibility. There are cartilaginous cancers called chondrosarcomas, so this one remains a myth.

- Ostriches hide their heads in the sand when they see danger. FALSE

This myth emanates from long ago, when Pliny the Elder (23–79 AD) wrote: "Imagine, when they have thrust their head and neck into a bush, that the whole of their body is concealed."[24] It may look as if the birds are hiding their heads in the sand, but in fact, when hiding from predators, they lie down, allowing them to blend in with their surroundings.

Instead of being clueless, ostriches have very acute hearing and vision, so they are able to detect predators from long distances. They are flightless birds, so this disadvantage is made up for by their excellent senses and rapid speed (up to 40 miles per hour). They fight using their strong legs and are anything but cowardly or stupid, contrary to popular belief.

Ostriches have been around for more than 120 million years, according to the fossil records. To have lasted so long, their strategies must work well. They are the fastest runners of all two-legged animals and have strong wings to hit back predators and to defend themselves.

- There is love at first sight. TRUE

Visual cues are perhaps the strongest stimulus for attraction in animal society. Females will choose males based on two factors: 1) their attractiveness, and 2) their resources within animal social systems. Attraction is the first attribute that a female bird, for example, will judge to select a mate. How much an individual's physical features are symmetrical determines attractiveness in most cases; the more symmetrical, the more attractive. Analysis of gene sequences show it is a sign of genetic vigor. In barn swallows, an experiment was done to make males appear asymmetrical. In all cases, females did not choose males who had wings that appeared different from each other. These visual cues are evidence for love at first sight in animals, but what about in humans?

A sizeable body of evidence suggests that human attractiveness is based on not one particular quality but on a general sense of symmetry. Symmetry or equality on both sides of the face and body are more important than any other quality in judging attractiveness. Studies of computer-analyzed faces with varying levels of symmetry were condusted with to a large number of males and females. Results showed that both groups chose faces that were more symmetrical as attractive.[25] This would make sense, evolutionarily, because genetic quality is linked to symmetry, and choice is based on what is best for one's offspring. Females will choose the most symmetrical male because that will be best for their young.

Culture plays a role in defining specific criteria for attractiveness. Research shows that certain qualities are ranked as attractive cross-culturally; symmetry; body fat distribution in females, musculature in males; smooth skin, good teeth, and uniform gait. As discussed in chapter 6, "The Media," the average American is exposed to about three thousand marketing messages per day. Body alterations in the quest for physical beauty are as old as history, with Egyptians using cosmetics during their First Dynasty (3100–2907 BCE). Hairstyles, corsets, body weight, and body piercing and tattooing trends have changed throughout human history. Scars have been viewed as masculine and a mark of courage in many societies. Tattoos were depicted on drawings and carved figures from Europe, ancient Egypt, and Japan many thousands of years ago. Body art gained prominence in modern Western society in the upper classes in the early nineteenth century but lost favor due to stories of disease spreading as a result of unsanitary tattoo practices. Lower classes then adopted the practices to show group affiliation. Tattooing has been used to denote classification: Indian tattoos demarked caste, Polynesians used marks for showing marital status, the Nazis marked groups from SS to concentration camp prisoners, and U.S. gangs show group membership. Tattooing has been firmly established in societies and continues to grow in popularity in the United States.[26]

These examples show the effect of society on determining physical beauty, but there are always consistent themes across cultures. Researchers asked people of different racial classifications to rate the attractiveness of Asian, Hispanic, black, and white women in a series of photographs. They found a high correlation in responses among all racial groups, ranking the same women attractive regardless of race.[27] This indicates that there is more than just culture dictating attraction.

Of course, the first part of the myth is about love. Defining love in the animal kingdom is difficult. Is there true love, or is it cold and calculating? Animal attraction can be measured and predicted based on symmetry. Love is more difficult to quantify. It is true that there is a bond between people and similar parallels between other animal species. For example, grief in greylag geese was well documented by Konrad Lorenz, father of ethology (animal behavior), who studied many of their behaviors. Geese will act very sad and depressed when losing a mate; this certainly shows dedication.

In most animal examples (and perhaps human as well), there is a predictable process to mating and to love. Choices are based on either attraction or resources. Both are tangible and measureable many animal experiments have shown this to be true. On a spiritual level of thought, this discovery is disconcerting. Perhaps religion and/or a belief in a higher power are an attempt to subjugate these innate genes dictating such a calculus.

Mate choices are made to benefit the young: choosing a mate with more symmetry might lead to better quality of offspring, or making the choice based on resources might give the young a better chance at life. One might say that it is for the sake of the children that choices are made in animal social systems. Still, it seems so cold to be able to predict who will mate with whom and for how long.

Consider the psychological model known as *matching hypothesis*, which often accurately predicts the matching of adolescents with one another based on their social standing in high school and their physical attractiveness. Is there a difference in animal societies? No. The male dung fly, *sepsis cynipsea*, will mate with females based on its strength and weight. Weaker, smaller males will remain on a female longer than larger males because smaller males have little chance at getting another female. In fact, a model called *optimal foraging theory* can correctly predict how long a male dung fly will remain on a female based on its weight. This relationship can also be applied to many amphibian species, wherein weaker males remain for such a long time atop females that drowning is a leading cause of frog deaths. If these are examples of love at first sight, I am not sure I feel comfortable classifying this myth as true; however, technically speaking, the stimulus for love is indeed derived from "first sight."

- Animals are kinder than humans. FALSE

Animal social systems cannot survive long without individuals helping one another. Cooperation is a key element in the success of any society, but what drives us to help one another? Is there true altruism, or are all animals, including humans, selfish? The answer is that humans and animals are both very selfish and cooperate based only on principles to benefit their own genes.

When helping is observed in nature, it is often between kin. Kin selection is the pressure on individuals to help one another because they share genes. For example, your mother, father, brother, sister, and children are 50 percent identical to you, which means that by helping them, you are actually helping your own genes. Your aunt, uncle, nephew, and niece share 25 percent of the same genes, so you are still inclined to look out for them, but less so than for the first tier. While it appears to be a cold and calculating view of human and animal kindness, such an assumption can predict most animal behaviors. Family is important in human society. You take your girlfriend or boyfriend to "meet the parents," nepotism is rampant at workplaces, and inheritance of assets usually moves down family lines (where there's a will . . . there are relatives).

Some animals help each other when they are not related. It is pleasant to imagine panda bears and prairie dogs as cuddly and caring of each other's well-being, but it is not so. Self-sacrificing is seen under unusual circumstances, and there is often a reason behind it. Vampire bats, *Desmodus rotundus*, are an example of a species with members who appear to help one another in an ostensibly altruistic manner. They are very small mammals weighing between fifteen to fifty grams (0.5 to 1.7 ounces), native to South America, and they feed on fresh blood. They are communal roosters, leaving their community in the night to forage for food. Vampire bats make a quick and painless bite on their victim, feeding for up to half an hour (about 20 milliliters of blood). They have a voracious appetite and produce an anticoagulant in their saliva to keep the blood flowing.[28] When returning to the roost they regurgitate blood to one another, whether kin or not, to help their fellow bat. It appears to be altruism, but upon closer look, helping behavior is linked to something more selfish, a "you scratch my back, I'll scratch yours" sort of agreement. Missing a meal is deadly for such a small creature, with a proportion of bats always missing a meal at any one night's feeding. Regurgitating blood pays off. At the next feeding time, a blood meal might be needed instead. This system is termed *reciprocal altruism* and depends upon members not cheating and mechanisms to detect cheaters; otherwise, the whole setup would fall apart.[29]

What about animal kindness to one another and toward humans? Dogs love their masters, but is it to gain food and shelter? Cats lick their owners, but is it for the tasty salt and urea on their skin? Strong bonds form between animals and humans and parents and children, but what is the motivation, genetically?

Konrad Lorenz (1903–1989), the father of ethology (animal behavior), and the current Richard Dawkins offer opposing viewpoints in the study of human behavior and motivation. In the selfish gene hypothesis, Dawkins argues that all life (plants, animals, and bacteria) is extremely selfish by nature. He claims that our genes puppet our behavior and that society should "try to teach generosity and altruism, because we are born selfish."[30] Lorenz argues that it is not human nature but society that leads to animal selfishness, particularly in humans; humans and animals are basically good, but society corrupts.[31] He points to examples suggesting that animal social systems demonstrate restraint in their aggression, but that our human inability to control violence may be due to our crowded capitalist system and is unnatural.

Lorenz looks to examples of animal social systems that cooperate for the common good of the group. Insect social systems such as the class *Hymenoptera* (ants, bees, and termites) exhibit cooperative systems. Sterile castes and a system in which an individual shows altruistic behaviors characterize eusociality. A bee will sting an enemy to save the group in a seemingly selfless act. This act occurs despite the fact that the bee's organs will rip out of its body along with the stinger and that within a few hours the bee dies. Worker ants are sterile and instead serve a queen master. All of this points to selflessness on the surface. Should humans take an example of how to treat our fellow citizens from the *Hymenopterans*?

Dawkins argues that the behaviors are solely based on an individual's desire to pass genes onto the next generation in what he terms the *selfish gene hypothesis*. This perspective views humans as shells that hold a very controlling genetic material that functions with a singular goal: reproducing itself. Social helping is explained in terms of haplodiploidy, whereby the queen gives parthenogenic (virgin) birth to all of the males in the colony. The males are thus haploid (containing half the full amount of DNA) and are all identical to one another. Thus, after genetic calculations, it is shown that females in the colony are 75 percent related to each other. This is because their father contributes all of his genes to his children. The children are exactly identical as far as their father's genes are concerned, and differences are due only to the mother's genes.

Thus, the insects' motivation (their genes' motivation) is what drives behaviors. In terms of helping, these types of organisms are more likely to help one another than others because of their high degree of relatedness. It pays to help each other because they are actually helping 75 percent of themselves (since eusocial insect females are 75 percent identical). For the most part, humans are at maximum 50 percent related, excluding identical twins.

There is actually a mathematical equation defining who should help who based on their genetics. In accordance with Maynard and Smith's equation to determine when helping behavior should occur, it is determined that eusocial creatures should help each other. The more related genes help one another. Natural laws of genetics may have more influence on human society than we would care to admit.

- Homosexuality is a result of parenting and early childhood environment. FALSE

Behavioral genetics is the branch of biology that examines the genetic basis of behavior. Studies indicate that sexuality (sexual drive and behavior) varies across a continuum in most animal societies, from asexual to hypersexual, with most individuals falling somewhere in between.[32] Most research supports a genetic basis of sexuality and very little influence of environmental factors. In monozygotic (identical) twins, a concordance (correlation measure) rate of 36.9 percent was observed for multiple sex partners and a 65.8 percent concordance rate for homosexuality.[33] Biological bases for sexuality lie in two factors: 1) the activity of the medial preoptic area of the brain (MPOA), and 2) the DRD4 dopamine receptor gene. The more dopamine activity in the MPOA area and the greater the number of DRD4 receptors, the higher the sexuality rates in humans and rats.[34]

The range in sexual drives and behaviors would appear to make sense evolutionarily. Hypersexuality may appear favorable for enhancing one's reproductive success (more offspring with more partners), but quality also counts. Thus, female sexuality is frowned upon in many cultures; genetically, it leads to a poorer quality of offspring. Consider that fertilization in many animals is enhanced by the presence of a seminal plug after a male ejaculates. If another partner enters, the plug is dislodged and this next partner is also able to father the child. In promiscuity, the final partner is equally (or even more) likely to father the child as compared with the first partner. Usually the last partner is weaker, older, and has poorer quality genes than the first. Thus, in animal systems, hypersexuality and promiscuity is selected against. Experimental evidence shows that hypersexual behavior in rats leads to decreased reproductive success for the female.[35]

At the other extreme, asexuality is observed in about 1 percent of the population, with such individuals having a lack of interest in sex altogether. However, the genes for this condition persist because lack of sexual attraction does not mean lack of sexual behavior. One would also expect homosexuality to be selected against as it does not lead to offspring in nature. However, pedigree (family history) analysis shows that homosexual individuals have higher rates of fertility than heterosexuals. One might assume that homosexual genes promote child bearing and thus perpetuate in the population.

Another hypothesis as to why "gay" genes remain in our gene pool is through kin selection. If people do not have their own children, they are more likely to help their nephews and nieces (kin), who are 25 percent identical to them, thus perpetuating their own genes.[36] While environment plays a lesser role in sexuality, it may be a contributing factor. However, the overwhelming evidence for a genetic basis of homosexuality and other sexual behaviors makes the myth false.

- Inbreeding is bad. FALSE

While inbreeding between close relatives causes genetic problems, a certain amount has been shown to produce healthier children. In a study of Iceland's family history lineage, marriage between third and fourth cousins produced the most and healthiest children over the past one thousand years.[37] Outbreeding with someone too different from your own genotype may lead to health problems in children. One such example cited was the Rh factor, which is a set of proteins on red blood cells that need to match for a healthy baby. If the mother, for example, is Rh negative and the father is Rh positive, then the second child who is Rh positive (from the father) will be attacked by the maternal immune system. Presently, RhoGAM is a treatment given to expecting mothers to prevent mismatched blood from causing an issue, but in our history, such a match was disfavored. Thus, there is an optimal level of inbreeding for reproductive success.

Inbreeding between closer relatives is known to cause health problems in children of such relationships and is frowned upon by every known culture in history. Laws banning incest are well ingrained in our culture. Inbreeding depression is the main culprit. It leads to the exaggeration of deleterious recessive genes in inbred offspring. When close relatives mate, both are from a lineage that may have bad genes that are usually covered up by dominant genes. Relatives have the same bad genes in common (which are generally recessive), so inbreeding has greater chances of revealing these recessives. Examples may include sickle cell anemia, cystic fibrosis, or even cancer. However, third and fourth cousins have only about 1/256 to 1/512 genes in common with one another, so the chances of such unions revealing recessive genes is quite low.

- It is warmer on coasts than inland. TRUE (usually, not always)

Coastal areas are surrounded by water and have more modified temperature changes. Water is held together by hydrogen bonds, which makes water a "sticky" substance. This is why an ant can get trapped in a water drop; hydrogen bonds attach all over the ant's body, making it difficult for the ant to move. Similarly, hydrogen bonds absorb and release energy to moderate temperature changes. When it becomes colder, water releases heat to the coastal areas, and when it becomes warmer, water molecules absorb heat from the air. As a result, in warm times, it is initially cooler on coasts than inland. The large ocean bodies filled with water at the coasts are lacking (by definition) in inland regions like the Midwest and Plains states, resulting in much wider temperature variation. Highs and lows in those regions are not moderated by water's hydrogen bonding potential, leading to colder winters and warmer summers.

- Alcohol makes you urinate. TRUE

The kidneys control how much urine is excreted from the body. Through regulating the amount of salt and other dissolved ions in the blood, kidney cells (called *nephrons*) reabsorb needed chemicals and secrete those the body does not want or need. Urea, an ammonia product, is dangerous as it builds up and may lead to death when the kidney fails.

Nephrons are our adaptation to living on land. Those cells retain the water needed for life while removing the toxic nitrogen-containing chemicals such as ammonia and urea. Many biochemical reactions produce nitrogen as a waste product. We need to remove it while, at the same time, conserving water. The kidneys perform this unique function.

Three hormones help control urine output by reabsorbing its water into the body: aldosterone, angiotensin II, and antidiuretic hormone. Alcohol is a depressant—it makes the hormones controlling the kidney slower and less effective. This is why "breaking the seal" is not a myth. People will stimulate their need to go to the bathroom because their urinary system slows from alcohol.

- Dogs do not sweat. FALSE

Dogs sweat through the pads on their feet; however, it is true that most of their heat loss is not due to sweating but to panting. When dogs pant they lose the hotter water molecules from their saliva, thus reducing the temperature of their mouths. Blood vessels in their heads and ears also dilate to bring more blood to the surface and lose heat. Humans also lose heat in this way, with faces turning red from vasodilation. A close look at a dog's tongue and lips will also show redness from dilation in the heat.

Without sweat glands throughout its body, a dog is at a disadvantage in regulating its temperature—not to mention that it has a thick coat of hair. Thus, misting or spraying a dog to create a layer of water molecules to evaporate would best be advised to cool dogs under hot conditions.

- Dark skin evolved because it protects from skin cancer. FALSE

It is true that melanin (skin pigment) causes darker skin tones and protects the genetic material of the skin cells from damage and thus developing into cancer. However, skin cancer usually strikes people after they have had children and passed their genes on to the next generation. Thus, darker skin would not develop in response to sunlight as a result of skin cancer.

Instead, research shows that sunlight depletes the skin of needed folic acid and vitamin B. Both of these are very important for a developing fetus, and lack of folic acid is associated with a host of birth defects, including spina bifida, Down syndrome, and autism. While folic acid is taken in easily in modern-day diets, in past society food was limited. Being able to retain more folic acid due to having darker skin was an advantage that flourished in sunnier, warmer latitudes and regions.

Alternatively, lighter skin colors developed in areas with less sunlight. Sunlight stimulates conversion of cholesterol in the skin into vitamin D, which is essential for strong bones and teeth. Present society has vitamin D in abundance (e.g., fortified milk) but again, prior eras lacked such nutrient access. This advantage is thought to have propelled the evolution of lighter skin colors.

Skin color is a trait due to only ten different genes out of millions in our cells. Surprisingly, this set of DNA is the main determinant of race and a divider of peoples in human societies. Skin color has been an obsession for humans since the ancients and before. Cleopatra painted her upper eyelids blue and her lower eyelids green. Ancient Greeks used ceruse for whitening their faces, and medieval people used red fucus to color lips. Both ceruse and fucus contain toxic lead components. Many skin lighteners and colorations used throughout history were toxic and must have had drastic health consequences for users.[38]

- You should not pick up bird eggs and place them back in their nests. FALSE

It is always best not to disturb natural processes, but this is a myth. Mother birds do not pick up the scent of humans on their baby bird eggs. Mothers have not been shown in observations to abandon their nests upon human scent detection. In fact, there are many examples of imitation eggs placed by one species in the nests of another. The mimic egg is raised by the host mother, and she is unable to detect that it is an imposter most of the time. Mimicry is a form of parasitism because it takes energy to raise young; the host mother takes on all of the responsibility while the other mother is free. There is an inability in birds to detect a mimic, human scents, and even the overall number of eggs in the nest.

- Dollar bill drop: A person cannot catch a dollar bill when it is dropped in between fingers. TRUE

When a dollar bill is dropped in between a person's fingers (thumb and middle fingers), with fingers around the face picture on the bill, the person will not be able to catch the bill. The rules require that the person dropping the bill drop it at a random time and that the catcher keep the hand stationary and not lunge forward toward the bill. The catcher's nervous system is always too slow to catch the bill.

The test is a measure of a person's reaction time, which is defined as the time it takes for a person to react to a stimulus. The stimulus is the drop of the bill. The reaction is the movement of fingers to catch the bill. A nerve message is sent from the brain after seeing the drop to the fingers to move them. There are too many nerve cells and too many gaps between the nerve cells (called synapses) for a person to ever be able to catch the bill. By random chance, one in one thousand people catch the bill, but usually the catcher is too slow. It is an excellent bar trick that can win some money in bets but also a black eye!

MEDICAL MYTHS

Myths most important to our survival are those that relate to and may even threaten our own health. Medical myths are pervasive in our culture because the educational structure is unable to address misconceptions. Chapter 6, "The Media," addressed the role of the media-educational complex in inhibiting science literacy. This contributes to a fomenting of myths in medicine because the media attracts more customers with hyped-up stories about health. The media-educational complex is most interested in medical myths because medicine is quite complex and media sound bites present pieces of interesting information. The purpose of this section is to complete the picture on several myths that have been perpetuated. Some are vital for human health while others are not so serious.

- Men over seventy-five should not be screened for prostate cancer. FALSE

This myth emerges because of the latest recommendation by the AMA (American Medical Association). Studies show that to prevent one death from prostate cancer, more than one thousand men would need to be screened. Several studies taken together show no statistical difference between the survival rates of those screened and those unscreened for prostate cancer. There are enough side effects from false positives and treatments that make the two groups statistically the same in outcomes. It is also true that if a man lives long enough, his chance of eventually getting prostate cancer equals 100 percent. Consider some serious questions upon taking the advice of the AMA: Is the recommendation valid, or is it merely expediency to limit testing and medical expense? Suppose you are the person who the test saves? Is a seventy-five-year-old ready to do nothing and merely die of prostate cancer?

The answers depend on the condition of the patient; if someone is on death's doorstep at seventy-five, then treatment may do more harm than good. Also, many prostate cancers are very slow growing, so many men with prostate cancer die of something else and not because of it. But one cannot be sure what type of cancer it is without biopsy and treatment. Statistics are powerful and show that screening does not extend life expectancy, which is the reason the AMA made its recommendation.

Screening with a PSA (prostate-specific antigen test) measures the amount of proteins given off by prostate cancer cells. At a certain level (2.5 to 4 and above) the number is elevated enough to call for a biopsy. A good doctor performs a digital examination to check if the prostate is rough or smooth: if rough, it is more likely to be cancerous.

Symptoms, patient family history, and other factors should also be considered before simply not testing people over seventy-five. This is probably a reason many men still get the test regardless of the AMA's advice. Screening catches cancers early and prevents some deaths; men aged seventy-five may live ten to thirty more years. I feel the AMA's advice is a myth because knowledge is power. I would want to know what the PSA test says about my own health. This is a rare moment when I make a recommendation against statistical advice, using commonsense reasoning instead.

- Screening for ovarian cancer and breast cancer is unnecessary unless there is a family history. FALSE

The BRCA 1 and BRCA 2 genes have a high correlation with breast and ovarian cancers. The U.S. Preventive Services Task Force, along with the American Cancer Society, recently announced that screening for these genes is unnecessary unless one has a family history of these cancers. The side effects of treatment and false positives from the test, it was determined, cause undue harm and overall do not benefit the average patient.

This recommendation is quite disturbing, and I place it in a medical myth category. Breast cancer strikes one in eight women, and ovarian cancer has a survival rate of less than 40 percent after five years. These are serious threats to women's health, and early detection dramatically increases the chances of survival to roughly 90 percent. Recommending cancer screenings based only on family history is ill advised because there are many gaps in one's history of disease: consider that many cancers are recessive in origin and skip generations. Do we really know what Aunt Martha died from in 1956? One in three people will die from cancer, so cancer is in *everyone's* family, not just some people's.

Screening for ovarian cancer specifically uses the CA-125 blood test. It is true that in a study of over seventy-eight thousand women, deaths from ovarian cancer did not differ between groups screened and unscreened. Also, 10 percent had false positives, and some women had an ovary unnecessarily removed. However, when screenings were included with a transvaginal ultrasound, more cancers were correctly detected. Vigilance is the answer to survival of any cancer, and early detection is our best defense. Statistics are being used to make decisions for the masses, but health is an individual choice.

Thus, in self-health interest, the test is better recommended for all women after forty years of age.

- The three-second rule: You can eat food picked up after it falls to the floor (without ill effects from germs) if it has been less than three seconds. FALSE

As soon as an object hits a surface, microbes attach. Research shows that one square centimeter of any random surface contains $10^8 = 100{,}000{,}000$ bacteria. The large number of microbes on any surface is destroyed by antibodies in our saliva, acidity in our stomachs, and our ever-present set of immune cells. That said, I do not recommend eating any food from the floor, even after one second, because the instant it hits, there is always the chance that the wrong, resistant virus was picked up and is now ingested.

- Alcohol is OK as long as you eat a healthy diet. FALSE

Alcohol's effects on the liver are the main problem for the heavy drinker. The liver breaks down toxic substances that are taken in by the body. Alcohol, in the form of ethanol (CH_3CH_2OH), is broken down by the liver to form acetaldehyde (CH_3CHO). Acetaldehyde (the good guy) stimulates the release of adrenaline-like chemicals that affect the brain activity associated with pleasure. So when you next hear someone say they want a drink, interrupt them and say, "No, you actually want a glass of acetaldehyde." This is sure to win you new friends! Acetaldehyde eventually breaks down to form carbon dioxide and water vapor, which we breathe out.

Now, for a description of the problems: During normal breakdown of food, the human body uses three chemical pathways to break it down into useable energy. First, a set of reactions called *glycolysis* make the food into pyruvic acid, yielding a small amount of energy and making NAD (nicotinamide adenine dinucleotide) into $NADH_2$, a high-energy molecule. Second, the pyruvic acid is shuttled into a cycle of chemicals known as the *Krebs cycle* to make more NAD into $NADH_2$. Both of these occur in the cytoplasm of the cell. The third step in getting energy from food is called the *electron transport chain*, which occurs in the mitochondria with the main purpose of converting the $NADH_2$ into useable energy. Alcohol slows down the first two steps (glycolysis and Krebs cycle) but greatly increases the third step (electron transport).

The problem that alcohol causes for these normal processes is a result of the extra hydrogens from the ethanol. Notice that two hydrogens are removed from the ethanol to form acetaldehyde. The loss of hydrogens (with their associated electrons) is termed *oxidation*, and we say that ethanol is therefore oxidized. The hydrogens need to go somewhere, so they attach to $NADH_2$.

The extra hydrogens are the bad guys; they are the culprits in liver disease. Why? Alcoholic hydrogens occupy the NAD, which would otherwise be used for glycolysis and the Krebs cycle. Instead, with NAD no longer available, food (proteins, carbohydrates, and fat) in the liver is not broken down but all turns into fat. Food does not go through the three steps (glycolysis, Krebs cycle, and electron transport) and sits in the liver without being metabolized. Mainly the $NADH_2$ from ethanol is shunted into the electron transport chain, leaving less opportunity to break down other food to be digested. Fats accumulate in the liver cells (called a fatty liver), and the cells die due to malfunctioning in this strange situation. The dead liver cells trigger an inflammation called *alcoholic hepatitis*. More and more liver cells die in this inflammation in which the body is attacking itself. The scarring is known as cirrhosis and is the ninth leading cause of death in the United States.

The evidence for this mechanism is in the abnormal histology (study of tissues) of liver tissue. Alcoholic livers have enlarged mitochondria (a part of the cell used in making energy) because of the large amount of electron transport occurring with the extra $NADH_2$. Another part of the cell, the endoplasmic reticulum, coats the excess fat onto proteins and also enlarges in alcoholic livers, illustrating the effects of increased fat deposits in alcoholic livers.

While the effects of alcohol on the liver are dangerous, alcohol is correlated with numerous other physical manifestations. Long-term usage effects are high blood pressure; heart disease; a weakened immunity; cancers of the esophagus, mouth, and liver; obesity; and muscle loss. Short-term effects are, of course, the hangover. Each of these effects has explanations at the molecular level and is worthy of supporting the argument against excess alcohol usage.

You may be thinking, "For all of this to happen it must take a long time. Thank goodness I have time to tone it down." You probably can predict what I'm going to report . . . yes, bad news. In a study conducted by Lieber and colleagues in 1985 at the Bronx Veterans Administration Hospital and the Mount Sinai School of Medicine in New York City, in a very short time (eighteen days) of heavy drinking (six ten-ounce drinks of eighty-six proof per day) an eightfold increase in fat deposits in the liver was seen.[39] These subjects were human volunteers fed a high-protein, low-fat diet, and still the results were dismal for the liver. The myth of eating a good diet to protect the body from alcohol's effects is not supported.

The controversy that underlies the alcoholic diet is not whether or not the diet is a good idea. Instead, it is the question every alcohol admirer asks himself or herself that elicits debate: "Why does a substance that gives humans such pleasure and yet so much pain even exist?" Who or what created such a duality? For ethanol to give cerebral pleasure and yet create so many hardships, one wonders why it is so . . . and why another substance giving only the good effects was not discovered.

In fact, why is it that the foods and/or beverages people like the most are also biologically destructive? Is it because there is a God who is cruel and likes to see humans tempted? Could it be that life on Earth is a test and that the temptations of alcohol are yet another; for some people it is more difficult than for others to pass up (perhaps because of different levels of desire for the substance)? Could alcohol be a randomly discovered substance that creates pleasure, and someday humans will find a way to rid the negative effects? The argumentation that emerges from the philosophical ramifications of the existence of alcohol can shake and realign our understanding of human purpose and design.

- It is better to break a bone than damage ligaments, cartilage, and tendons. TRUE

Bone is able to heal within six to eight weeks, generally, if bones are set properly. The structure of bone tissue is very vascular, meaning that it has many blood vessels available to send in healing cells. Bone is also cheap to build (it does not take much energy from the body), with calcium salts and collagen fibers of protein able to be manufactured easily.

Conversely, ligaments and tendons are very energy expensive to build because they are very orderly materials. They are also poorly vascularized, with little ability to heal. Tendons connect bones to muscles and, upon hyperextension (with blood vessels), tend to tear easily if against their grain of collagen fibers. If a tendon is torn, for example, more than halfway through, it generally requires surgery. Ligaments, which connect bones to other bones, have the elasticity of taffy—about 4 percent pliability—and then they break. The worst of the damage is to cartilage, which is avascular, meaning that no healing is possible through blood vessels. Cartilage has very poor healing capacity, which is why it often requires replacement or scraping to help patients. About 80 percent of cartilage is water, so shrinking back of the material sometimes happens to help the healing process in situations such as disc herniation and bulges.

- A broken bone is stronger than the original. TRUE

Bones remodel according to the forces placed upon them. This is a principle in anatomy known as Wolff's law, which purports that osteoblasts build new bone tissue when pressure is applied or when there is a break in the bone, and osteoclasts break down bone when bone is remodeled. There is a continuous cycling between osteoblast and osteoclast activity. Every six to ten years all of the compact (hard) bone in our bodies is replaced, and about every three to five years spongy (soft) bone is replaced.[40] It is strange to think that all of our bones are being replaced, without us knowing it, throughout our lives.

Bones overcompensate in areas of breaks, and a broken bone is indeed often stronger and thicker than originally. If a long bone of a femur is cut lengthwise, there is an area that is thickest near the center. It is also coincidence that this area is also the region most likely to break, based on Euler's buckling equation (an engineering formula that predicts mathematically the spot on a tube that is most apt to buckle). Euler's equation from which this information is derived is $Pcr = pi^2 El/L^2$, with L equal to the length of the tube, E the elasticity of the substance, and P_{cr} the critical load amount to cause buckling—but the details are unimportant at this point. The formula is not known to our osteoblasts and osteoclasts, but still the body thickens at the section of the femur most likely to buckle in order to prevent a potential break.

- Scaring a person, eating sugar, breathing into a bag, and drinking a glass of water are the best ways to stop hiccups. FALSE

Imagine hiccupping for sixty-eight years. Charles Osborne, an Iowa farmer, hiccupped from 1922 until 1991 (when he died) every few seconds, except while he was sleeping. It would ruin your life, and a cure would be priceless. He had the longest attack of hiccups on record. Hiccups are common, though, and extended forms of hiccups can result from damage to the phrenic nerve (it originates from the fourth cervical nerve plexus, connecting the neck to the diaphragm). Scaring a person, eating sugar, breathing into a bag, and drinking a glass of water are *not* the best ways to stop hiccups. They are not very effective, especially not in chronic cases such as Charles Osborne's.

The best cure for hiccups was discovered by accident. A sixty-year-old man with acute pancreatitis was admitted to the hospital. He was immediately fitted with a nasogastric tube, bringing with it an onset of hiccups lasting two days. A variety of treatments for hiccups was attempted with no success. During a routine rectal examination the hiccups abruptly stopped. When the hiccups resumed again a few hours later, the rectal exam was repeated, and the hiccups stopped again.[41] As you may deduce, Charles Osborne did not live to see his cure; what a cruel irony.

- House dust is old skin. TRUE

Most dust on furniture and in houses and closets derives from skin either from humans or insects. Insect skin is mostly shed or is dead matter consisting of chitin protein, the most abundant protein in the animal kingdom. Human skin continually grows and is shed. The outer layer of our skin, the stratum corneum, is about twenty to thirty cell layers thick and sheds at a continual rate throughout our lifetimes. In fact, if we did not shed this skin, we would have an extra layer three feet wide all around our bodies by the time we hit eighty years old. Our arms and legs would be six feet in diameter, covered with stratum corneum! These layers are filled with keratin protein and are similar to chitin, forming dust particles everywhere indoors.

- Humans have five senses. FALSE

The five senses include sight, sound, touch, taste, and smell. However, there are many other pieces of data detected by the human body. Internally, special proteins known as *receptors* pick up information about conditions inside and outside of the body that are sent to the brain for interpretation.

For example, proprioceptors are found in our muscles, tendons, and joints to send information about body position and balance to the brain. Damage to muscles and joint areas can lead to balance problems. Some medicines, such as cholesterol-reducing statins, may have side effects of muscle and proprioception damage, leading to poor coordination.

Nociceptors sense pain, which is not included in the original five senses. Pain sensations are very much receptor-driven but also have a psychogenic component. There is a mind-body connection that may enhance or inhibit pain, according to many sources. Evidence shows that relaxation and calming thoughts and environments can reduce nociception and even eliminate pain. There is also a cultural component to pain, with some cultures exhibiting more pain reactions than others. Areas of medicine and research are dedicated to pain studies because pain is so prevalent in our human condition. Almost one-third of all people suffer from some form of chronic pain.

- You can get warts from toads or frogs. FALSE

There has never been evidence of toads or frogs as being vectors of warts. Warts are caused by a virus that invades skin cells and causes a growth. A wart is actually a benign tumor, which does not spread and is not a malignant cancer. While warts are not medically life threatening, they are troublesome and do spread from person to person if the virus enters an open or delicate area of the skin.

In fact, you cannot generally get warts from other species, such as frogs. Viruses are largely species specific, which means that they only spread between organisms of the same species. A cat with a viral infection does not spread its sickness to humans despite close contact. However, viruses do spread between people. If a small cut or surface on the skin is open, a virus could be picked up, and they are at times difficult to remove.

Additionally, when a toad appears warty, it does not mean that it has warts; it simply looks roughened. Viruses occasionally mutate and spread between species. We believe that the HIV virus that causes AIDS was spread from apes to humans at some point in its mutation. However, for the most part, viruses of other species do not spread to humans.

- Shaving causes hair to grow back thicker. FALSE

Hair grows at roughly one-half inch per month. Hair follicles each grow a hair based on genetic and hormonal influences. When hair is shaven it only appears thicker because the newly growing hair is stubble and shorter hair looks more solid; thus, the hair appears thicker. However, the same number of hair follicles remain before and after shaving. The number of hair shafts emerging from these follicles also remains the same.

It is true that continuous irritation or rubbing can stimulate hair follicle growth. Evidence of over-the-shoulder straps leading to more hair on a particular side of the body is documented. However, it is with a good amount of time and rubbing that some growth is stimulated. Rubbing is not a solution to hair loss because the amount of stimulation to the hair follicles is insignificant.

- Dental X-rays cause brain tumors. FALSE

News reports recently made the claim "Dental X-rays cause an increased risk for meningioma, the most commonly diagnosed brain tumor in the U.S.," and other headlines expressed similar findings. However, in a rush to report the results of a study by the Yale School of Medicine, the media neglected to report the numbers.

The study looked at 1,433 people diagnosed with intracranial meningioma, a tumor that forms in between the tissues lining the brain. It compared the different types of dental scans the patients had undergone in their lifetime and found that those who had yearly bitewing X-rays were at between a 40 percent and 90 percent higher risk of being diagnosed with a brain tumor. Upon closer analysis of the number, however, it would mean only an increase in lifetime risk of intracranial meningioma from 15 out of every 10,000 people to 22 in 10,000 people; clearly an insignificant difference.[42]

Limiting exposure to ionizing radiation in forms such as X-rays and CT scans is a very important point nonetheless. CT scans are a series of X-rays that image body parts in a three-dimensional manner, giving perhaps some of the best medical images. The radiation is over two hundred times that of a regular chest X-ray and should be used sparingly. Overall lifetime risks significantly rise for cancer as the number of CT scans increases for a person.

The thyroid is a gland in the neck that is particularly prone to the damaging effects of radiation. Cancer and malfunction is evidenced in patients exposed to even small amounts of radiation. Thus, a thyroid shield is recommended for any person receiving a dental X-ray, regardless of the type of imaging test. The dentist or dental hygienist is required by law to provide this shield but will rarely offer it; it must be requested by the patient. There is no harm nor is there difficulty or discomfort in using the shield. It is just another step that the dental clinic does not offer readily because it is more work. Dental X-rays do not significantly increase the risk of brain tumors, but any extra radiation is bad radiation.

- Flu shots cause the flu. FALSE

Flu shots are derived from weakened (attenuated) or dead viruses or pieces of viruses and pose no harm to the patient. There are odd cases of a vaccine with an attenuated virus strengthening, but it is very rare. These are merely myths for almost all cases. The danger in the omnipresence of this myth is that the case exists (e.g., polio or influenza) in which a vaccine actually caused the disease for which it was used to prevent. However, the myth should be debunked based on sheer statistics and facts: 1) the flu vaccine does not contain a live or weakened virus, so it is impossible to get the flu from the flu shot today; 2) 225,000 people are hospitalized each year due to influenza and between five thousand and fifty thousand will die depending upon the year and severity of the virus. The flu vaccine is recommended, especially for at-risk patients such as seniors.

- Cold weather causes the common cold. FALSE

The common cold is caused by the rhinovirus, which is spread from person-to-person contact or through airborne means. Cold weather in no way causes viruses to be born and infect humans. However, cold weather does drive more people indoors and increases the population density in particular spaces at any one time—in the mall, in schools, or in dorms. Thus, during times of cold weather, common colds (and other sicknesses such as influenza) are more frequent. The density of people increases transmission of the rhinovirus because more contacts are possible for spreading the illness.

While a runny nose is a part of any cold day, it runs to clean out the respiratory system and is a functional part of immunity. However, cold weather has been shown in studies to suppress the immune system due to the physical stress of the weather. Suppressed immunity gives the rhinovirus the opportunity to gain a foothold in our bodies and invade cells.

- A tan is healthy. FALSE

Skin darkens when ultraviolet (UV) light hits melanocytes on the skin's surface, which then produces melanin. Melanin is the pigment (or paint) of the skin that gives protection to the skin from light and creates darker skin tones. It was believed that tan skin protects from the effects of UV light. This is true and is probably from this factoid where the myth originates.

However, UV light is shown to be associated with many unfortunate and disease-causing results. Skin has an abundant amount of protein ropes holding it together, called collagen. These are white fibers that give support by revisiting tension (pulling) forces and compression (pushing inward) forces. This way the skin has "give" and strength. Collagen is found in numerous areas, including the kidney, blood vessels, joints, and discs of the back. It is everywhere, giving structure to the human body.

When UV light hits collagen, it forms cross-linkages and becomes brittle. When moved, the collagen cross-linkages break, and wrinkling and sagging result. That effect is only on appearance, but UV light also links closely with skin cancer.

The rate of skin cancer incidence has increased markedly, with the very dangerous malignant melanomas occurring for 1 person in 1,500 in 1930, to 1 in 250 in 1981, to 1 in 87 in 1996, to 1 in 79 in 1999. This pattern is likely to increase by about 7 percent more cases per year. Skin color and social status have been associated since ancient society. In the time of Cleopatra, the desire for lighter skin was associated with upper-class social status. By the 1920s, the attitude had changed, with media advertisements pushing tanning lotions and darker skin tones. Americans spend almost $400 million on suntan lotions and other cosmetic products that help them tan.[43] However, a tan is nothing more than destroyed collagen and changing cells into cancer.

SPACE AND MOTION MYTHS

The most important part of the space program may not be its ability to contact life or discover materials in outer space for human use. It may well be the motivation that it draws for people to become interested in science. Everyone has gazed at the stars in awe and wonder. Harnessing that intrinsic excitement about the universe and our relation to it brought many children into science in the past and keeps many adults in continual fascination.

This section examines many facets of space and motion to show continuity between human and material law. We are linked to our universe in so many ways: we are composed of atoms made from stardust carried out here from the Big Bang era, our chemical nature dictates our health and, fitness, and as seen in medical and nutrition myths, humans are subject to the laws of nature that are many times predictable and humbling.

- The moon has a dark side. FALSE

Every section of the moon is irradiated by sunlight at some point in a day. The side of the moon facing the Earth always faces the Earth. We never see the other side because of Newton's universal law of gravitational attraction, which states that all bodies attract to each other based on certain variables: $F = G(m_1 \times m_2 / \text{distance}^2)$, whereby the masses of the bodies attracting each other contribute to the force of attraction, but the distance weakens that force.

The extreme pull between the Earth and moon prevents the moon from rotating one face away from the Earth; this is known as *tidal locking*. The moon still rotates on its axis, but it is in synchrony with the Earth's rotation, so that one side of the moon always faces Earth. The dark side of the moon is more properly called the "far side." The far side of the moon was first photographed by a Russian space probe, *Luna*, in 1959. It was observed with human eyes first in 1968 by Apollo astronauts and photographed numerous times in our space missions. The far side contains many craters and bumps, described by astronauts as appearing like a sandbox—messy and full of holes. A large crater known as the South Pole–Aitken basin, about 1,500 miles wide, has been suggested as useful for future exploration to obtain samples and to give greater understanding of the composition of the moon.

- The moon came from the Earth. TRUE

Most atoms come from stars long dead—from the fusion of original hydrogen atoms derived during the Big Bang and still combining and dividing to create new materials. Space, the universe, and Earth's origins have common atoms; however, when comparing lunar and Earth soil samples, it appears that the moon derived from the Earth at some point in Earth's early history. Lunar rocks are compositionally similar to Earth's mantle; both are composed of similar silicon isotopes, indicating that the moon came from the Earth. This is called the Giant Impact Theory, which asserts that the moon came out of the Earth after a giant meteor hit the planet (see figure 8.4).

According to this theory, a Mars-sized developing protoplanet called Theia orbited the sun alongside Earth. Theia was too small to remain in orbit and varied its

Figure 8.4. Giant impact theory. Source: Hemera/ThinkStock.

distances from Earth as it orbited the sun. Eventually, about 4.53 billion years ago this instability led to its final crash into Earth at what is believed to be a sixty-degree oblique angle, thrusting portions of each into space. This material gathered to form the moon. Evidence to date supports this claim, with silicon isotopes in similar proportions compared to Earth's mantle.

- There is no gravity in space. FALSE

For the same reasons that the moon orbits the Earth and both orbit the sun, there is also gravity in space. There is pull from the nearest object, depending on where in space one is located. Based on the law of gravitational attraction, there is a force seizing hold of any object in space, creating a pull of gravity. The object with a larger mass always has the greater pull.

In space, astronauts only appear to be weightless. In fact, the force of gravity is only 10 percent less when a spaceship is 250 miles away in outer space. Astronauts are being pulled in their spaceship back to Earth by gravity, except that they are moving sideways to negate the pull downward. It is similar to a carnival ride in which the cart spins sideways and one feels weightless. Gravity is everywhere in outer space; some areas with bodies far away have little gravity, and yet black holes have so much mass and pull that they even trap light, making the region appear "black."

• Humans *cause* earthquakes. TRUE

The phenomenon known as *induced seismicity*, whereby human activity causes miniearthquakes that may lead to larger quakes, is well documented in geology research. The U.S. Geological Survey (USGS) reported many seismic events as a result of gas drilling and filling of hydroelectric dams. The USGS reported a series of forty-three small earthquakes (largest magnitude, 2.8) that were linked to hydraulic fracturing (fracking), which involved the injection of high-pressure water and chemicals into the ground to break apart shale rock formations.

Some research has also shown that surface loading (dam water pressure) may contribute to earthquakes. Millions of tons of water filling a body create pressure on the land below, leading to greater chances of landslides and seismic events. It is questionable whether a large quake (6.3 magnitude) in India in 1967 was linked to the filling of a dam nearby. Some studies argue that surface loading results in seismic activity,[44] but results are still tentative. However, smaller events that link to hydropower dams are well documented.

Earthquakes are a result of large land movements. First, Earth's crust has large plates that move regularly since the formation of the Earth 4.5 billion years ago. Geologists refer to this movement process as *plate tectonics*. Earth's plates are depicted in figure 8.5.

Second, movements emanate from energy deep in the inner layers of Earth. A temperature differential between the hot Earth's core and cooler temperatures toward the Earth's surface leads convection currents. This drives vertical circular movement of mantle fluid. As a result, plates on the surface either move away from each other or toward one another. Movement of plates creates zones of tectonic activity, which are areas where the plates collide, and this creates an earthquake. Locations such as California, Japan, Indonesia, and Alaska are along the edges of these plates, as shown in figure 8.5. When the energy of the plates is released during the movements, shaking and damage in the land above occurs. The energy (heat) deep within the Earth comes from the historical formation of the solar system and the energy of the Big Bang.

• Life violates the Second Law of Thermodynamics. FALSE

This misconception is pushed by organizations that want to show that the order life creates is unique. It is true that life's complexity and arrangement is special and a wondrous creation. However, there is no violation of physical laws. The Second Law of Thermodynamics states that all systems tend toward entropy (randomness). A child's room becomes messier; a set of books on a shelf gets out of arrangement; and one's hair gets disheveled. All of these examples are natural processes and occur spontaneously as time goes on.

Conversely, it takes work to accomplish the opposite effect of these actions. A child cleans up a room, a librarian rearranges books on a shelf, and one combs one's hair—all take work and effort. Each action requires work and effort and would not occur on its own. The second example violates the Second Law only on the surface.

Figure 8.5.   Major tectonic plates of the earth. Source: iStockPhoto/ThinkStock.

The energy derived to do the acts given emerges from energy of the sun. Life is ordered and maintained in a way that acts against randomness. But it is the energy of the sun, trapped by photosynthesis and harnessed by energy reactions, that drives the order of life.

The sun is continuously breaking down helium and other atoms to produce the energy needed for life on Earth. Its fusion and fission reactions are natural increases in randomness on the sun's surface but are used by life on Earth. Thus, we are not violating the Second Law but are a subset of the reactions that conform to it. Randomness is battled against by living systems. Disease is often a result of the breakdown of order created: a tendon tears, a plaque roughens an artery, or a disc bulges to interrupt the order that life created. It is true that life is a secondary effect of the increasing randomness of the sun, but the order life creates is still marvelous. Ultimately, the sun is predicted to burn out in 4.5 billion years, ceasing the energy that drives life processes. We still have a great deal of time until then.

- Lightning never strikes twice. FALSE

Lightning is *more* likely to strike the same place twice. Electrical energy is attracted to metal and any material with moving electrons. Taller objects are more likely to be hit than shorter. There is often an attribute, such as inthese examples cited, that make a spot more likely to be hit multiple times. If a location was hit once, then it is probably that type of attractive spot.

The Empire State Building gets hit with lightning twenty-five times a year because it is tall and has electron-inducing materials. Staying atop the rod on the Empire State Building in a thunderstorm because it has already been hit is ill advised.

- Dropping a penny from the Empire State Building will kill someone. FALSE

The aerodynamics of a penny would not allow a speed fast enough to kill a pedestrian. A penny is about two grams, and it should reach a terminal velocity of 65 miles per hour when hitting the ground, based on a gravitational pull of 9.8 meter per second squared. At this speed and with this mass, it would feel only like a bite, not something leading to serious injury. Of course, with wind drafts and the fact that lower levels of the Empire State Building would block the penny from hitting the ground directly, this myth is easily debunked.

- Solid wire is more useful than stranded. FALSE

Strands increase the flexibility of wire and prevent kinks, breaks, and fatigue. Strands contribute to increasing the total elasticity of a tube, without which many electrical wiring projects could not be accomplished. Stranded wire is more expensive to build than solid, but the benefits are substantial. In fact, living systems invest in these structures all of the time.

Stranded wires are found throughout living systems. For example, our muscles are arranged as stranded fascicles, with wrapping upon wrapping of muscle fiber tubes around insulation (connective tissues). Thus, muscles are better able to withstand pressure and pull from joint and bone movements. Young's modulus gives the amount of pull a material can withstand, and it shows that stranded tubes are much better than solid materials in resisting forces.

- Seasons are due to the distance of the Earth from the sun. FALSE

The sun and Earth do move closer together at certain points. The Earth's orbit is not a perfect circle around the sun; instead, it is an ellipse. Elliptic orbiting does lead to some increases in temperature when the Earth nears the sun, but the change in minimal.

Seasons are due to the tilt of the Earth along its axis and the angle the sunlight makes with the hemisphere facing the sun. If a hemisphere is facing the sun more directly, that hemisphere will be in a warmer season, such as summer. If the hemisphere faces away from the sun, it will be in a colder season, such as winter.

Seasonality is due to the angle of solar energy and not proximity to the sun's surface. It is true that solar flares and solar activity influence Earth's temperatures. Opponents of global climate theory, contending that nonhuman factors affect climate, use solar data. Sociopolitical controversy clouds the debate on this topic, unfortunately.

Temperature changes due to the sun's proximity (Earth's ellipsoid orbit) are not measureable, however, due to complicating factors of seasonal variations and topographical (mountains, oceans) temperature conflating the measurements. Theoretically, warming on earth does take place, but it is predicted to be minimal and not cause seasonality.

- Saturn's rings are solid. FALSE

Saturn has thousands of planetary rings. These rings are primarily composed of water-ice pieces. Ice pieces are typically about one meter apart and range in size from specks to chunks the size of a house. Gravity causes the ice pieces to collect into ring shapes around the planet.

Galileo was the first to observe Saturn's rings in 1610. It was, however, Christiaan Huygens who first described them as a disk surrounding the planet.

- Meteorites are hot when they land. FALSE

The chunks of rocks and ice that form meteorites have been traveling through space in extremely cold temperatures (around 450 degrees below 0°F) for millions of years. Thus, they are very cold when they begin their descent through Earth's atmosphere.

As they reach the atmosphere the outside begins to heat up, forming crust. At the same time this outer crust begins to strip off, in turn removing some heat. Falling through the atmosphere takes only seconds, so for large rocks, only the outside has time to be heated up. The rocky inside is a poor conductor of heat, leaving the center of the meteorite cool. The outer heat from the fall may be entirely lost by the moment of impact on Earth's surface, depending on the size and composition.

CONCLUSIONS ON SCIENCE MYTHOLOGY

The myths in each section are a product of misinformation and misconceptions; in the absence of an effective science curriculum, national standards and measurements are not used consistently across states for educating young people in science. Thus, science literacy is threatened. Obviously, some of the myths supported or debunked are not life-threatening, but some are. The role of the scientific community in ensuring accurate and effective transmission of information to the public is addressed in the next chapter.

9

What Are Scientists' Responsibilities?

Mundus vult decipi (The world wants to be deceived).

—Walter Kaufmann

CONSUMERS OF SCIENCE

In a world easily taken in by pseudoscience, myths, and just plain bad science research, it is easy to accept that people want to be deceived. This chapter explores the role of the science community in combating the concept of *mundus vult decipi*. Through education, motivation, and research, an individual can become a user of science instead of a victim of misinformation. Whether a person is facing a medical procedure or contemplating the veins on a garden plant, she or he requires scientific thinking and needs to be informed.

Our science community expands beyond the science teacher, researcher, and medical doctor. Included first is the consumer of science and technology—the public. People are the customers who determine how science ultimately affects society. Many of the myths and pseudoscience presented in the previous chapters were based on information that is inaccessible to the public, and this is why the information needs to be addressed in this book.

What is the responsibility of consumers of science? To be able to make decisions and appreciate nature for their own sakes and for the sake of their loved ones. Without knowing the facts behind many of the misconceptions that arise when presented with scientific information, the consumer has a high probability of making the wrong decisions. A person needs to be his or her own advocate when navigating the medical edifice to reach the best decisions on his or her health. Only through understanding the science can they best judge the consequences for their decisions.

161

The number of medical errors in the United States has skyrocketed in the past decade. It is estimated that 32,500 patients die each year as a result of preventable medical errors. Over three-fourths of these deaths are due to prescription-related mistakes.[1] It is a worldwide problem, with the World Health Organization estimating that up to 10 percent of all patients are affected by medical errors. Operating on the wrong kidney, nicking a vein or nerve, and amputating the wrong leg are all medical horrors. Some cases may be unavoidable mishaps, and others could have been prevented. It is patients' responsibilities to minimize their own risk by weighing the evidence and helping their own case.

If you are getting hip surgery, it would not hurt to remind the staff on which side to operate. It is important to write down questions, take notes, and research your own diagnosis when interacting with a doctor. This is all advice easily given by many sources but difficult to follow. Doctors are in a rush and under financial constraints. In order to turn a profit, a general practitioner can spend no longer than twelve minutes with each patient. Thus, while the medical community advocates asking lots of questions, many doctors resent the time and effort of answering. It is with this in mind that reliance on authority has led many patients to disappointment and much worse. Only through education and a scientific way of thinking can the public become more effective consumers of science. This chapter aims to give advice to scientists and nonscientists to prevent and combat integrity issues in science.

SCIENTIFIC INTEGRITY

The information available to consumers must be valid and based on set standards set forth by the scientific community. Without honest reporting of data, accurate statistics, and proper study design, the information available to the public is useless and dangerous. As discussed in chapter 3, "Tools Scientists Use," the many methods of creating and evaluating scientific information allow the public access to a great deal of knowledge. In reality, tools can be misused, and science is not always to be trusted. Ethical issues are one of the greatest threats to science, and there are guidelines to combating these problems.

HONESTY

Honesty is the first chapter of the book of wisdom.

—Thomas Jefferson

Honesty is the most important aspect of science research and its resulting information. Without real data the decisions made by both scientists and consumers of science are flawed. Science is a community, and researchers depend on shared information that is honestly reported and verified. If one researcher finds that seismic activity can be detected using a particular instrument but its precision is misreported, all of the information derived from that instrument is useless. Money and time are lost, but the decisions made could be deadly. An earthquake could have been predicted but was not because of the faulty equipment, for example.

There are a host of dishonest practices that occur within the science community. The reality is that we do not know how much dishonesty exists because such behaviors are covered up. Research requires money and political support to advance. There is an incentive to publish results and draw further money. Whenever money is involved, the temptation follows alongside.

Often, the pressure to produce positive findings and desired results leads researchers to unethical behavior. Grant money or the threat of job security may be factors, but it is has been shown to be a recently increasing trend. The number of retractions due to fraud or misinformation in scientific journals increased in the past decade alone.

Scientific misconduct can manifest as either fraud or plagiarism. Fraud is more serious because it constitutes the falsification or manipulation of data to produce a desired (but false) result. I have seen among colleagues, both as a student and professor, examples of fraud in science research and teaching. This is not to point fingers or make accusations but to heighten the awareness in people to not merely trust authority. How unfortunate that science research should be questioned at such a base level. However, fraud has existed as long as science has and before. Many institutions ignore fraud in their research divisions due to fear of reputation damage.[2] This chapter aims to clarify for consumers and scientists the systemic aspects of fraud so they may better combat it

An example of outright fraud is found in the case of Stephen Breuning, who falsified the results of studies on drug therapy for severely mentally disabled children. He falsely recommended stimulant treatment over tranquilizers, thus endangering the lives of thousands of children. The treatment he recommended had adverse effects on children. The results were published with a singular goal of improving his academic standing without performing the prerequisite work.[3]

As is often the case, it takes the courage of a colleague to expose misdoings. A fellow scientist, Robert Sprague, discovered that Breuning had made fabrications and reported the misconduct to Breuning's grant funding agency. Whistle-blowing is very dangerous for the reporting scientist like Robert Sprague. There are political ramifications, and in science, as in many areas, people protect each other and try to bring down the whistle-blower. In fact, Sprague found himself under investigation and political attack for reporting Breuning. It is the ethical responsibility of scientists to report misconduct, and it is a shame that so much is allowed to flourish. Reporting should be done carefully and discreetly with senior staff and institutional support, but also extrainstitutional oversight.[4] Documentation is an important legal factor in reporting unethical science, and only the written word is valued in a court of law.

PLAGIARISM

Plagiarism is another form of dishonesty in academic areas. It is defined as the use of someone else's words or ideas without giving that person credit. Generally, when an idea is taken from another work, the source is cited so that the idea can be traced back. Original meaning is important to establish the facts behind research. Also, credit is given as a matter of respect, and it establishes the history of the ideas. Intellectual property laws make it clear that any person's original thoughts are their property and are to be given proper deference. In this book, information is cited for these purposes.

Of course, there is a judgment call involved in any decision to cite an idea. Some ideas, like "bodies fall to Earth due to gravitational pull," should not be cited to Sir Isaac Newton. The idea is common knowledge, and it should be assumed that we all know it. On the other hand, judging what "everyone" knows is sometimes unclear, and citing can become cumbersome to readers. As a golden rule, when in doubt, cite and cite some more.

There are a host of other kinds of scientific misconduct, as reported by the National Academy of Sciences.[5] The examples that follow provide a guideline to the scope of misconduct that may be used by scientists to harm the public in their own interests. Through the addressing of these potentialities, consumers of science may become better able to detect misinformation and myth. First, manipulation of public opinion on a particular issue occurs frequently when the results of a study are prematurely released without a full peer review. The peer review process is extensive and serves as a quality control check to science information. It will be discussed more fully in the next section.

The media have been shown, throughout this book, to report findings without sufficient evidence or partial evidence in order to gain ratings. The media-educational complex feeds off its use of technology to gain excitement from the public about certain science findings. Consider the following example: initial reports on cold fusion (which is a type of nuclear reaction occurring at low temperatures to yield large amounts of useable energy) claimed a breakthrough, and chemists B. Stanley Pons and Martin Fleischmann held a press conference to announce their findings prematurely. In this case it was the scientists' enthusiasm that fed the media. Pons and Fleischmann had not gone through the data enough to warrant a press conference. After about a year, other scientists could not replicate the research findings, and it was deemed that Pons and Fleischmann did not actually achieve cold fusion.[6]

Pons and Fleischmann are guilty of overenthusiasm and misjudgment. In contrast, many times scientists know how to use the science process properly but choose to mislead and misinform. Statistics and interpretation can be very misleading when an author knows how to do so. Some of the inappropriate uses of mathematics or statistical design to interpret data were discussed in chapter 3, "Tools Scientists Use." To recall, this kind of deception may lead to public health consequences, and the tobacco industry has been a notorious culprit. In the reporting of only certain mathematical analyses in the 1950s and 1960s, Big Tobacco downplayed the link between smoking and lung cancer. It is possible that such mathematical misinformation has led to one out of every nine people in the United States currently dying from a smoking-related cancer! In fact, the number of cancer-related deaths is expected to rise substantially in the developing nations by 2030 because of the misinformation that has spread within their culture about smoking.

Another form of misconduct in science is the failure to give access to information: proper and timely data reports, physical evidence, or procedural information and mathematical analyses. Keeping a timely and meticulous lab notebook is crucial in any scientific investigation and is a legal document to back any findings or process in science. In fact, the U.S. National Institutes of Health require that the written documentation of the findings contributing to any of their grants be kept for three years after publication.[7]

Publication of material is a serious part of a science academic's professional life. Through conferring proper authorship and/or proper credit for contributions to a published work, the scientists' efforts are recognized. Again, academic jealousies and competitiveness often obscure proper rewarding of contributions. In a sense, it is intellectual theft to omit crediting a particular contributor. In the graduate school world, there are many tales of this behavior taking place. Often, students become the servant of their advisor or dissertation chair and do not know or are too afraid to claim credit for their work. Some professors are fair and just, giving proper credit to whomever gave useful contributions, and others are not.

It is true that in some cases, the credit to be given is unclear: does a lab technician who performs only procedural tasks but does not contribute cognitively deserve authorship? Does a person who merely restates a hypothesis or previously known information deserve original credit? Answers to such questions are dependent upon the situation, but it is clear that credit needs to be given for science contributions to progress fairly.

Most people are unfamiliar with scientists as a community. In all of the cases of scientific integrity issues discussed above, it is always the character of the scientist conducting the investigation that determines the honesty of the work. The science community is very small (in one's own area of research), and news travels quickly if there are questionable issues of integrity. I remember a Johnny Cash song with the line "Bad new travels like wildfire, good news travels slow." Social pressure by science community members is stronger than one might think. The following section discusses the unique aspects of scientists and their value systems that help rein in ethical issues. If the conscience of an individual scientist fails, social pressure of the community often leads to the prevention of scientific misconduct.

PERSONAL RESPONSIBILITY

At the height of World War II, V-1 and V-2 rockets bombed England, killing thousands of civilians. Germany was the only nation to develop the use of rocketry, shocking the world with its scientific advances. After the war, Dr. Wernher von Braun, rocket developer for the Reich, was accused of being responsible for the many dead from his invention.

In response, von Braun once announced that "science does not have a moral dimension. It is like a knife. If you give it to a surgeon or a murderer, each will use it differently. Should the knife have not been developed?" This perspective was enough to allow von Braun residency in the United States and a career with NASA.

Our government not only overlooked Wernher von Braun's accountability for war deaths but eagerly "denazified" him and made von Braun a U.S. citizen. This scenario raises the very provocative question: Should scientists be responsible for the results of their discoveries? And what about political leanings and uses? If the United States had not found use in von Braun's scientific innovation, would he then have been placed on trial for war crimes?

Wernher von Braun laid the foundation for the development of modern rocketry, which, of course, is the basis of our military strength and advantage to this day. Do scientists have a responsibility for the uses of their research? Should STEM (science, technology engineering, and math) professionals be held accountable for the results of their developments?

We have discussed Wernher von Braun, but what about the developer of the automobile? In 1672, Ferdinand Verbiest, a Jesuit missionary priest in China, built the first steam-powered vehicle for the Chinese emperor. This was the first working steam-powered vehicle, or "automobile." Is this innocent priest, who built a toy for the Chinese emperor, responsible for the 1.8 million deaths per year in motor vehicle accidents? The work of any scientist may be used or abused by people after its development. Sometimes its evil side is obvious, and other times, unpredictable. It is up to each reader to determine the extent to which they personally believe scientists should be held legally and morally accountable for their work.

The developer of the atomic bomb in the Manhattan Project knew full well that his work would be used for destruction. Robert Oppenheimer, a nuclear physicist, commented upon viewing the first explosion of the weapon, "Now I am become Death, the destroyer of worlds" (a line from the Bhagavad Gita). Radioactivity emerged quickly in this development to change sociopolitical boundaries. Awe and fear accompanied the nuclear application of Albert Einstein's equation relating matter and energy, $E = mc^2$ (where c equals the speed of light). It has changed society in so many ways. Wars with nuclear weapons would destroy many more, nations use nuclear fuel for energy (30 percent of U.S. electricity is powered by nuclear power plants; in France, 80 percent), and politics uses nuclear weaponry as a form of deterrence. What are Einstein's and Oppenheimer's roles in changing the world? Do they have personal responsibility?

ETHICAL DILEMMAS

It is difficult to know the answers to these questions. To determine how to develop science, the effects on society are rarely considered. Science develops because there is always someone who will continue working on the knowledge. Humans have innate curiosity. Our role is to best predict and prepare for how science will change our lives.

A recent battery of prenatal genetics tests is being developed—over 1,500 minitests for all sorts of genetic diseases. Prenatal genetic testing, which screens developing embryos before birth, makes it possible to determine fetal genetic diseases. Right now, cystic fibrosis, neural tube defects, Huntington's disease, and sickle cell anemia are screened. The effects of prenatal testing on the future of many unborn fetuses are obvious. If a bad gene, which may lead to a birth defect, is detected, a decision to terminate a pregnancy is likely for many individuals. What if dwarfism, Alzheimer's disease, or depression were to be screened? What if a baby's future height, weight, eye color, IQ, or other vanity trait were to be determined for expecting parents? There is the potential to develop a society of "designer babies" and change the genetic-social stratification layers of society. In the 1990s movie *Gattaca*, a futuristic world is depicted that is getting closer and closer to becoming reality.

Consider the following scenario arising from prenatal genetic testing. Would you terminate a pregnancy if the test shows that a) the child would suffer from spinal muscular atrophy and die within nine months of birth; b) it would have cystic fibrosis and is expected to suffer lung problems and die by age twenty; c) it will have Huntington's disease, in which the person is healthy until age forty and then suffers progressive muscular weakness and dies by age fifty (many younger readers draw the line against termination here, because they think at forty, one has lived a full life already!); or d) it will get Alzheimer's disease at age sixty or suffer manic depression its whole life.

These ethical dilemmas should be uncomfortable for the reader. Evaluation of ethical choices takes reasoning and personal accountability. It is a question of when life begins but also of religious belief and personal situation. How simpler choices were before technology made this possible. Determining an unborn child's genetics is a very recent development. Modern science has it shortcomings, and, as discussed in chapter 3, "Tools Scientists Use," there is always the chance of error adding further difficulty to decision making.

If the baby were to be born minutes before you made a decision about its termination, would you consider killing the baby after it is born? What if the genetic testing shows that your baby will suffer horribly throughout its life? Most people would say "no" because it is against the law. Obviously, killing the newborn is illegal and is murder. But it is the legal system that sets the level of reasoning, and it is at a low level of authority.

In ancient to medieval times, long before prenatal testing, it was quite commonplace to abandon a newborn upon finding out about diseases. It was common to abandon a newborn if there were too many children and not enough food to eat. There was no genetic testing, and this practice was a social norm for screening infants. Clearly, human laws change with the discovery of new techniques such as genetic testing; how will our laws change once genetic testing becomes expanded?

OUR OWN BODIES

What right does the state or medical agencies have in controlling our own bodies? This, too, may change in the future. Will STEM developments advance enough to implant a chip into our skin to "help" us understand information better and make us smarter?

A best-selling book, *The Immortal Life of Henrietta Lacks*, by Rebecca Skloot, discusses a medical case in which Henrietta Lacks, whose cervical cancer cells were harvested and grown in 1951, dies from the disease but medical authorities used her cells in the lab for study. The cancer cells in Ms. Lacks were taken from her cervix by biopsy as part of her diagnosis, and in fact have been kept alive in vitro for the past sixty years. A characteristic of cancer is that its cells are immortal, and if they are given enough food they will live forever.

Cancer cells contain an enzyme known as *telomerase* that builds up on the ends of DNA to prevent them from ending their cell division. Senescence, or the aging process, is a result of cells not being able to be replaced when worn out. Telomerase allows these cells to keep dividing. So, is telomerase the fountain of youth? Unfortunately, numerous studies have shown that the addition of telomerase results in tumor production, the opposite of eternal youth.

Thus, cancer cells do not age or die as normal cells do. The cancer cells from Henrietta Lacks are called the HeLa cell line. These cells come from the same set of cells harvested from Henrietta Lacks when she was alive. Her cells have been vital for cancer research. Recently, they have shown the relationship between human papillomavirus and cervical cancer and have led scientists to develop a vaccine for cervical cancer.

The viral-cancer link has social consequences as well. The vaccine is now established for cervical cancer and is being marketed to nine- to fourteen-year-olds to prevent transmission of this type of cervical cancer. The research in HeLa cells gives estimates that roughly 10 percent of all cancers are viral in origin, perhaps more.

The major legal and ethical questions that arise from HeLa cells are whether or not a person "owns" their own cells. Was Henrietta Lacks wronged by having her cells used without her permission in 1951? Does it matter—she's dead anyway? Do her heirs have rights to her cells? There is litigation by her family members to gain the rights to those cells and the money that was gained from the research. Human laws often attempt to control scientific discoveries and the impacts on society, but how effective are those laws?

HUMAN VERSUS NATURAL LAW

Natural law does not conform to the will of humanity, society, or human laws. It does not change with sociopolitical or economic dynamics. Humans cannot control nature, but the attempt to do so is a strand running through our history.

Nature has tremendous power, and human society is unable to control most phenomena. Earthquakes, solar flares, tsunamis, and perhaps the nature of human behavior are part of the power of natural law. Max Otto (1876–1968), the science philosopher, once reflected upon this relationship: "The universe is run by natural forces and laws, not by moral laws . . . Humans live by moral laws. If [they] contradict, it will be human society that suffers the consequences."[8] Human society continually struggles against nature in ways that appear foolish to the onlooker. There were some successes for sure: wind power, heart surgery, farming techniques, and even nuclear fission are examples of scientific applications of human invention that better our lives. All too often, though, our efforts are thwarted, and we recognize weakness.

As humans with tremendous cerebral thought, we tend to think of ourselves as organisms high on the evolutionary chain. Our power and influence on Earth is to be respected. However, to what extent do we really have control over the planet?

Since Earth's origin, there have been continual shifts in the climate from the molten rock upon which heat killed all life to ice ages that probably led to the extinction of many species. World climate has changed through geologic time. Ice cores from the polar caps show these fluctuations in modern geologic time. The "Little Ice Age," an era of unusually cold weather that occurred between 1550 and 1850, after the medieval warm period, had significant impacts on society. It is believed that this cold was a result of changes in solar output and a volcanic eruption. Agriculture failed throughout the Western world, which brought about starvation, desertion of Greenland's colony population, and even cannibalism in Europe. It was during this time that killing children to conserve food supplies was so commonplace.

Clearly, there were no significant anthropogenic (human-related) atmospheric emissions at the time of the Little Ice Age. Humans did not contribute to the climate change of the Little Ice Age. That does not mean that we are not contributing to changing the world's climate today. Presently, we see a strong correlation between global temperature and atmospheric carbon dioxide levels. We see major shifts in climate and microclimate patterns throughout the world, from Antarctic ice melts to a warming trend in certain areas of North America. Industry remains a main contributor to carbon and sulfur emissions. Are these pollution emissions the cause of current global climate change? Do we as humans have the power and strength to willingly sway global temperatures? Can human law move the climate back toward a favorable direction, or are nature's machinations beyond our control?

Global climate change (global warming) has become a hot-button geopolitical issue wrought with human conflict between business and environmental interests. A search for the truth leads to political dynamics that are not pure science. Our purposes are not to take sides in this book but to encourage us to question and investigate freely. The essence of science requires this.

SCIENTISTS ARE THEIR OWN SOCIETY

The Big Bang Theory is a popular TV sitcom that depicts the lives of a small group of young scientists in California. Stereotypes abound, with nerdy science characters showing off their knowledge but getting into many predicaments in a world they do not understand. There are truths the show highlights, but there are also misconceptions about the nature of the scientific community.

Scientists comprise many unique groups bound together by a common interest with common education on how to view the world. Their weltanschauung is grounded in the scientific method and a way of thinking that is very different from that of the larger community. There are many fragmented metapopulations of scientists working deeply in their particular area of science interest. These are usually intelligent people looking to solve problems and discover new ideas. However, they are not knowledgeable about all areas of knowing; they often have deep specific knowledge of a particular domain.

In some studies, it has been reported that even scientists lack science literacy in areas other than their own. In a study of about two dozen geologists and physicists asked to explain the difference between DNA and RNA, only three could do it, and genetic material was an important aspect to all of their research.[9] The finding underscores the narrowing nature of science communities. Now consider if a group of scientists could recite a line from the "The Love Song of J. Alfred Prufrock" by T. S. Eliot . . . I suspect fewer of them could than compared with a group of literature majors. There are intellectuals who are scientists and scholars in many areas, and these are vital role models for society. Breadth of knowing should be a goal of STEM education because narrow scientists miss the "whole" picture of an investigation.

Science is a very collaborative process through which ideas are shared and knowledge is created. In fragmented societies, scientists communicate and contribute in a subculture very different from the dominant group. *The Big Bang Theory* sitcom has it right when it depicts a group of people cut off from the larger world because of their differences. The scientific community continually seeks a balance between communicating their research with the world and maintaining their unique structure. Oliver La Farge stated this of the scientific community in the following:

> Thus at the vital point of his life work [doing research] the scientist is cut off from communication with his fellowmen. Instead, he has the society of two, six, or twenty men and women who are working in his specialty, with whom he corresponds, whose letters he receives like a lover . . . in the keen pleasure of conclusions and findings compared, matched, checked against one another—the pure joy of being really understood.[10]

A set of norms exists to maintain integrity—the main responsibility of the scientific community to the public. Within this subsociety, a set of norms exists. John Ziman described these norms with the acronym CUDOS: communism, universalism, disinterestedness, originality, and skepticism. First, *communism* means that science is the property of everyone. Scientists share their information by publishing and presenting their data in public forums, and they should not hide their data. In this way, ideas are exchanged, built upon, and improved or debunked. It is a communal process.

However, in times of war and in cases of competition between companies, this rule has been violated. Nonetheless, science history is rich with examples of collaboration and progress: Sir Isaac Newton, the developer of mechanical physics, attributed his successes to this comradery in the comment, "If I have been farther it is by standing on the shoulders of giants."[11] As such, science should be shared with one another regardless of a scientist's religion, gender, nationality, or political persuasions. Ziman terms this *universalism*, meaning that national borders or classifications make no difference to the interpretation, use, or quality of scientific results.[12]

At the other end, a scientist's past accomplishments should have no bearing on his or her future results. A good reputation should not allow for a halo effect, in which anything one particular scientist publishes gets special treatment. In reality, it would be naive to suggest that favoritism and politics do not influence science inappropriately. Unfortunately, the battle against this unethical giant is difficult. Again, money and connections make certain figures more influential and their research more respected.

On the other hand, the system works well many times, with universalism trumping favoritism. To illustrate, Linus Pauling was a famous chemist who is known for elucidating the nature of chemical bonds and the structures of molecules, but he could not convince his contemporaries that large doses of vitamin C had medical benefits. The scientific community objectively evaluated his evidence and rejected his claim.[13] As discussed in the previous chapter, the link between vitamin C and cold prevention remains a myth.

Unlike politicians, who have a vested interest in outcomes such as elections, disinterest is another vital characteristic of the science community in upholding ethics. *Disinterest* means that the scientist should not have a personal interest in the outcome of the study. While professionally there is often pressure to produce a positive hypothesis or to pursue a certain direction of research by authorities, it is highly unethical for a scientist to choose politics over science. Thus there is an inherent mistrust of authorities and politicians within the scientific community. Carl Sagan stated this succinctly, warning that one of the greatest commandments of science is "Mistrust arguments from authorites."[14]

Science requires *originality* to advance, and merely restating old ideas or reformulating information is not enough to bring about new connections. Any useful dissertation has original thought and results, although many dissertations these days lack both. Finally, the fifth trait of the scientific community is an inherent *skepticism*. Skepticism is a theme of this book because it underlies the most important aspect of a scientist—individuality. The science community is skeptical, and the peer review process for judging work is intense and scrutinizing. The feedback can be very useful for the researcher viewing his or her own work through the eyes of another. Again, the process is laden with corruption. Often the reviewers have vested interests: they may be writing their own paper on the topic; they might know the author (despite anonymity) and jealousy may lead to a negative review; or the reviewers may simply give poor advice. Despite these flaws, science research is more readily modified and critically evaluated in an iterative process.[15] Few works, including this one, are published in the same form as their original composition. Ideas move back and forth between scholars to produce a better product. Critical evaluation can be tough for the scientist to listen to, but it is an essential part of the peer review process.

PEER REVIEW

The peer review process is a bulwark against scientific misconduct. An "invisible community" of colleagues judges new research anonymously. Journal articles are the primary conduit for new data announcements in the form of individual studies. The articles are refereed, meaning that peers with expertise in the same area of research judge them. Second, presentations are given at scientific conferences held by organizations within a scientific discipline. This is often a valuable first step in the submission of a journal article for publication. I always enjoy presenting because there can be useful feedback and discussion between a presenter and the audience. Third, books are written to incorporate and refine ideas derived from many sources, both one's own and others. Books are an integrative style of research, requiring the author to develop new syntheses and bring forth ideas that singular papers cannot.[16]

Scientists also gather online socially, formally, and regularly in an informal process of collaboration. The Internet allows for immediate electronic-transmission of information so that science occurs much more rapidly than in the past. As early as fifteen years ago this was not possible en masse. Informal interactions such as social events, job seeking, funding opportunities, and even gossiping comprise an important part of scientific information sharing that often is underestimated.

REWARDS

Competition for limited resources of grant money and publication space helps maintain quality and research standards. While it has led to unethical behaviors, competition drives progress. During the peer review process the peer reviewers are kept anonymous, and the author is likewise unknown to the peer reviewer. While this is not always the case, it helps minimize some effects of academic jealousies and differences. The reward system is based on the quality of the product and not politics.[17]

Individuality and independence are important aspects of a scientist. However, a scientist should care about the recognition of his or her ideas by their community. In this way, the work can be funded and furthered via collaboration with colleagues. Often, rejection and isolation hamper a perfectly good research agenda. Sometimes it is warranted, but often it is not; navigating politics is always difficult for the objective nature of the scientist.

BECOMING A SCIENTIST

Scientific interest begins in childhood as an innate interest in the natural world. The excitement of children's eyes when seeing their first firefly or learning about how the moon moves is contagious. My children augment my own passion about science as I see it through their excited eyes.

It is disconcerting that our interest diminishes so rapidly as we are exposed to science in the traditional classroom. Once in the education pipeline, fewer and fewer students remain interested in science as they progress, with over 85 percent of those originally interested in the elementary grades renouncing it later. This problem will be further discussed in chapter 11, "Getting People to Love Science." There are many attributes to becoming a good scientist—organization, patience, library skills, writing skills, and higher-level analytical reasoning—but the most important is motivation. Scientists need to have a freedom of thought different from that in many other areas. Many scientific wild-goose chases have resulted in discoveries of great import, such as radioactivity. These accidental discoveries might not have taken place without a science mind free to muse.

These descriptors make scientists a unique community, to an extent cut off from the larger society, but always sharing a similar weltanschauung of objectivity and process. Science people work very hard and long hours in a cognitive capacity known only to their subgroup. To the world, they are far from normal.

I do not wish to perpetuate myths about scientists or to group them as a monolith, but the shared attributes are what make science so special. An example will help illustrate the concept. In the 1940s, when General Leslie Groves assumed command of the Manhattan Project at Los Alamos, he was shocked at the work habits of the scientists. Instead of a focused group of intellectuals and scholars, he witnessed all types of strange social behaviors, but also a very social group: concerts, parties, and dances, and gossiping and neuroticism, along with odd and frivolous pursuits. Dr. Robert Oppenheimer, the head scientist at the time, explained to the general that this is how scientists worked under pressure.[18] Given the information in this chapter, it is no surprise . . . And yet, under these conditions the atomic bomb was produced to win World War II. An enjoyable time was had by all, with a group of scientists socializing during the very serious production of the atomic bomb.

III

SCIENCE

Threats or Compromises

10

Science Progress and Challenges to Science

An optimist is a person who sees a green light everywhere, while the pessimist sees only the red stoplight . . . The truly wise person is colorblind.

—Albert Schweitzer

A ROCKY ROAD TO SCIENCE

Argument and debate move society forward to create new knowledge. Without it, old ideas merely reinvent themselves. History is beset with conflicts whenever new ideas threatened the status quo. Whether by a discovery of fission or the creation of Facebook, society is rocked by a shifting of ingrained thoughts.

All existence has controversy inherent within it. People, ideas, philosophies, and science phenomena all have opposing sides. Consider the examples: light and dark, positive and negative, proton and electron, sun and moon, positive forces and negative forces, good and evil. Perhaps it is my argumentative side, but as much as nature has opposition, scientific study involves opposing sets of viewpoints.

Science is filled with dichotomies. A strange fact becomes obvious during a solar eclipse, as shown in figure 10.1. When the moon covers the sun in this event, the moon and the sun appear equal in size. Most people know that the sun is much larger than the moon: the diameter of the sun is 867,000 miles and the diameter of the moon is 2,155 miles; thus the sun is four hundred times larger than the moon. Then why do they appear to be the same size? Because the moon is also four hundred times closer than the sun. The moon is, on average, 238,857 miles from the Earth, while the sun is ninety-three million miles away. After calculations, the distance difference and the diameter difference are equal but opposite. While this is a freak

Figure 10.1.   Solar eclipse.
Source: iStockPhoto/Think-
Stock.

example of the many strange dichotomies in science, it is emblematic of the tensions and oppositions found embedded in science history.

In chapter 2, "Science Is Arguing," I emphasized the philosophical history of the ancients. Aristotle, Plato, and Socrates observed the world and developed very different views of how it operates. Science history continues with ideas that conflict with existing thought. Breakthroughs and changes occurr only when individuals pierce through the patterns of the times.

Louis Pasteur, founder of microbiology in the nineteenth century, knew the symptoms of rabies: fever, headache, tiredness, and drooling. He had a case in which a boy became infected, and Pasteur could not live with the prognosis: 100 percent chance of death. Pasteur injected the boy with a spinal cord mixture, taken from dead dogs, that he had been working on. After a series of injections, the boy slowly improved, and became stronger, and in the third month of recovery, Pasteur announced that the child was out of danger. Pasteur's patient was the first rabies victim to survive in human history.[1]

It takes a leader in thinking, in experimenting, and in individuality to make progress in science. History shows that these efforts are uncommon and often fought against by the dominant culture. Even George Patton, an American general during World War II, noted of his troops: "If everyone is thinking alike then somebody isn't thinking." Ways of thinking in science require individuality, as shown by Louis Pasteur.

Pasteur's discovery of the rabies vaccine demonstrates the innate human ability to overcome contemporary ways of thinking and to elicit change. His discovery took place in the nineteenth century but emerged from the period of skepticism borne of the scientific revolution. Europe rapidly advanced during the era through scientists' discoveries and methods that are still guiding science.

Scientific advances did occur before the scientific revolution. There were the Greek and Egyptian inventions of their alphabets; bronze and iron metallurgy; and glass, metal, and plastics development from Islamic society. Over 4000 years ago BP (before present), China's *Medicine Book of the Yellow Emperor* distinguished between twenty-eight different kinds of pulses and stated that the heart pumped blood.[2]

Throughout its history, science occurred at different rates during different periods. In some periods, science occurred at a punctuated and accelerated rate, rapidly advancing; in other times, it lulled and even moved backward in thinking. Humans advanced by using a variety of materials available to them. First, natural materials such as elephant tusks, deer antlers, and stone tools were the technology of innovation. Cultures thrived with these discoveries. Then, chemicals in nature were mixed and modified, giving rise to the smelting of metals and the cross-linking of rubbers.[3] Modern industrialism and transportation emerged to create a population boom and to propel the Western world scientifically and technologically. We are entering an age of science in which there is the synthesis of new materials at the molecular level in the form of nanotechnology and biomimicry. Scales of materials are getting much smaller, leading to very new ways to deal with scientific questions. Could a chip be implanted to attack cancer at the molecular levels? This is being developed in labs as we speak. Will a cartilage mimic be created to solve back and neck pain forever by inserting a rapid-growth biomimic of cartilage into the joint space to replace worn out intervertebral discs or knees? This would circumvent the need for hip, back, and knee surgery, which are primarily caused by the deterioration of cartilage covers on the ends of bones.

One theme emerges, though—science cannot be stopped, which is a main focus of this chapter.

CHANGING VIEWPOINTS

After the discovery of disagreement in the ancient world, a move to embrace skepticism characterized the scientific revolution. In the sixteenth century, the age of exploration led to communication with other cultures and other forms of science. Thus, Europe changed with incoming products and ideas. The invention of the printing press, dissemination of books, and contact through colonization built new ways of thinking. For example, new navigation methods were sought for overseas exploration. This required new instrumentation, such as telescopes, and encouraged research in mathematics and astronomy to improve travel. In Europe, the growth of medieval universities formed a society of philosopher-scientists to challenge the larger communities. In these ways, Western philosophy grew to change the observation-based Aristotelian science. Systematic data collection and a search for newer explanations prevailed in Europe. A movement to isolating variables and objectively analyzing problems took hold in an era known as the Scientific Revolution (1540–1690).

THE COPERNICAN REVOLUTION

Italian scientist Galileo Galilei (1564–1642) was put on trial for claiming that the sun was the center of the universe and not the Earth (which was the opinion of the

Figure 10.2.   Galileo Galilei. Source: Hemera/ThinkStock.

times). Galileo was sentenced to house arrest in Florence for the rest of his life and forced to reject his "false opinion." His portrait is shown in figure 10.2.

This trial is a main example of the tension between authority and individualism. The trial is a symbol of the scientific revolution. Skepticism eventually won out, and Galileo was shown to be correct. But the scientific revolution was not an incident but a conversion of thought. Challenges to scientific thinking still occur. The purpose of this chapter is to trace the changing ideas about science through history and to show the potentials for the future.

HISTORICAL IMPACTS

John Donne complained in a poem in 1611 that "new philosophy calls all in doubt." This quote represents a public, much like ours today, frustrated with changing information given by the scientific community.

Galileo's ideas were based upon the work of Nicolaus Copernicus (1473–1543). He published a mathematical formulation long before Galileo's time, stating that the Earth was not at the center of the universe. Copernicus believed that the sun was at the center and that "what appears to be the motion of the sun is in truth a motion of the earth."[4] Placement of the sun at the center of the universe, termed the *heliocentric model*, deemphasizes the importance of the Earth. Thus, any believer in the heliocentric model was suspect by religious and government authorities of the time. In fact, Giordano Bruno (1548–1600), an Italian monk, continued work on Copernicus's ideas, publicly teaching that the universe is infinite and has no surfaces and no end. Bruno was burned at the stake in a rage to keep the status quo.

PLANETARY MOTION AND FEATURES

This did not keep astronomy from developing. Johannes Kepler (1571–1630), a German scholar, used mathematics to describe planetary movement. His three laws of planets are: 1) planets move in ellipses around the sun; 2) a planet's speed varies with its distance from the sun; and 3) planetary movement can be expressed mathematically. Kepler recognized the order in the way the physical world operates. Kepler and Galileo communicated with each other, and Kepler urged Galileo to continue his work. Both were concerned for their safety.

Galileo used the telescope to show that the moon's surface was imperfect, with craters, holes, and ridges. This countered Aristotle's views that heavenly bodies were perfect in form and function. Because of Galileo and Kepler, the universe was not a perfect creation but had strange patterns and imperfections. Religious views were also shaken; if the planets were supposed to be heaven, then how could they not be perfect? Galileo was forced to renounce his views, making concessions on points to get his book, *Dialogue Concerning the Two Chief World Systems*, published. The controversy is obvious, but the solution to suppress science did not work.

NEWTONIAN PHYSICS

Sire Isaac Newton (1642–1727) developed much of modern mechanical physics, demonstrating laws of motion that well describe much movement in the physical world. He formed the following laws of motion: inertia, acceleration, and action/

reaction. Newton also developed calculus, a mathematical area studying rates of change, and developed the relationship formula for attraction between two bodies based on their mass and distances, as discussed in chapter 8, "Debunking Science Myths: Separating Fact from Fluff." Engineering science relies heavily upon Newton's ideas.

Until Albert Einstein (1879–1955), who modified mechanical physics to include space and time as variables, Newtonian physics was the only accepted way of understanding the universe.

REVOLUTION SPREADS

Medicine, anatomy, and chemistry experienced major changes in the way they were practiced. Investigations in these areas occurred in unorthodox ways for the times. For example, Paracelsus (1493–1541), a Swiss alchemist-physician, rejected Aristotle's idea that water, fire, earth, and air made up all matter. Instead, he looked at how chemicals in the body work and what remedies would help people. He administered arsenic and mercury to cure various ills; while they are poisons and probably caused great harm to his patients, he laid the foundation for modern pharmacology. His work showed that humans are composed of a variety of chemicals in different combinations. Paracelsus's work led to the birth of modern medical drug research.

In the study of anatomy, Andreas Vesalius (1514–1564) wrote a 1543 treatise, "On the Fabric of the Human Body," which used observations from his human dissection work. Such work was unethical at the time because the body was considered sacred and it was not legal to dissect it. However, he would secretly pick up bodies on the roadside after they had died and dig up human corpses for observation. The social pressure against his work eventually caused Vesalius to give up on his research. William Harvey (1578–1657) extended the work on cadavers, observing various cardiovascular principles. Harvey primarily studied animals, making comparative observations.[5]

On an even smaller level of observation, Anton van Leeuwenhoek (1632–1723) observed specimens under the newly discovered microscope. In the 1670s, he described "little animal or animalcules" he observed in water samples taken from lakes. These are what we now know to be microorganisms—what an explosion of diversity! That so many creatures lived in a world we could not see with the naked eye was unimaginable at the time.

Complexity within these "little creatures" led to a drive to find out about a whole new invisible world. The birth of microbiology can be attributed to Leeuwenhoek. Unfortunately, smaller structures within the cell were not observed until the discovery of the electron microscope in the 1930s.

On an even smaller scale, Robert Boyle (1627–1691) was interested in the chemicals composing the human body. He laid the foundations of modern chemistry. By extending the work of Paracelsus, Boyle relied on experimentation and new instruments to search for basic elements of matter. He also discovered a law describing

Table 10.1.   Key Dates in the Scientific Revolution

1543	Copernicus's heliocentric model was published.
1543	Vesalius published *On the Fabric of the Human Body.*
1609–1619	Kepler developed Three Laws of Planetary Motion.
1633	Galileo's trial.
1673	Van Leeuwenhoek observed animalcules.
1675	Boyle's Laws were discovered.
1687	Issac Newton published *Principia.*

the nature of temperature, pressure, and volume, termed today as Boyle's Law. As the temperature rises on an object (e.g., a can of soda), its pressure inside increases; hence, it explodes upon opening.

These great thinkers of the scientific revolution built the foundations upon which modern science is built. Skepticism prevailed, and new ways of thinking continue today. A few key dates are shown in table 10.1.

THE ENLIGHTENMENT

The scientific revolution sparked an age of reason known as the Enlightenment (1733–1789) that brought scientific thinking into nonscience domains. Methods of science were used in studying people and society, money transfer, religion, and moral law. The modern disciplines began in this period: psychology, sociology, anthropology, economics, and theology and ethics. The new intellectual movements were accessible to the public because of the increased use of the printing press, developed by Johannes Gutenberg in 1440.

There was the dissemination of about twenty-five thousand copies of the *Encyclopédie* (1751–1772), with contributions from hundreds of leading great thinkers of the time such as Montesquieu (1689–1755), Voltaire (1694–1778), and Jean-Jaques Rousseau (1712–1778). The Enlightenment questioned all authority, reaching against the counterreformation of the Church and various political structures and economic systems.

The age of questioning was termed *Aufklarung,* or clarification. Enlightenment thinking fomented the American Revolution of 1776, beginning with the Declaration of Independence and the Bill of Rights. American founders Thomas Jefferson and Benjamin Franklin were very much influenced by European skepticism.

The era's achievements in free thinking sparked a movement toward political freedom and working against the monarchy, particularly the British. Much like in most eras, the end of the Enlightenment saw a restriction in thought. However, the spirit of and optimism about science continues today. Modern scientific breakthroughs would have been impossible without the free thinking developed during this period.

The era witnessed science developments in chemistry by Antoine Lavoisier and physics through building upon Newton's laws. Magnetism and electricity, atomic theory and chemical behavior, and longitudinal lines in mapping are a few examples of science progress during the Enlightenment. Science used rational and secular critical-thinking strategies to analyze problems.

BRIDGING TO THE TWENTY-FIRST CENTURY

Science in the seventeenth century laid the foundation for the collaboration necessary for further research in the "new" disciplines such as mechanical physics, chemistry, and psychology. The rise of scientific journals and societies during the Enlightenment enabled scientists to work together to form breakthroughs.

The growth of the modern university system allowed the organization of scientists into disciplines. In the first decade of the nineteenth century, German universities were first to offer natural science as a field of study, and schools in other nations soon followed. Universities advanced scientific knowledge by promoting original research and by training scientists and other professionals who would spread scientific thinking throughout society. Some examples follow: Dmitri Mendeleev (1834–1907), a Russian chemist, compiled a new chart of chemicals known as the periodic table; James Maxwell (1831–1879) analyzed the relationship between light, magnetism, and electricity, leading to modern electronics; Ivan Pavlov (1849–1936), a Russian scientist, studied the conditioned reflex in dogs; and in medicine, new tools such as the thermometer and stethoscope helped diagnose ailments by measuring human vital signs.[6]

The microscope enabled users to examine body tissue samples, and X-rays (1896) led to images of internal structures. The electrocardiograph (EKG; 1901) traced the electrical activity of the heart to diagnose cardiac disease. Louis Pasteur (1822–1895) explained how people caught infectious diseases through his experimentation called *germ theory*, discussed earlier in this chapter.

All of this led to improvements in medical treatments, including sterile techniques such as the use of antiseptics, hand washing, face masks, and rubber gloves during operations and patient care. Florence Nightingale, a nurse during the Crimean War (1853–1856), stressed cleanliness, fresh air, and discipline to combat human illness. By the turn of the nineteenth century, medicine was "professionalized" through the use of her recommendations.[7]

While much of eighteenth and nineteenth century science encompassed classification and observation as its focus, it led to paradigm shifts that changed science today. Charles Darwin's (1809–1882) theory of evolution in the late 1800s was based on patterned observations of life's characteristics. As shown in his *Descent of Man and Selection in Relation to Sex*, published in 1871, Darwin's ideas did not use experimental or chemical bases. Instead, Darwin used induction to figure out how life developed. Looking at birds' beaks and the similar traits of different creatures, Darwin described a change in species over long periods of time due to the environment. A finch had a beak in a certain shape, for example, because it was best adapted for certain types of fruits.

Evidence for evolution emerged slowly with data coming from chemistry and with energy and matter more specifically understood. The twentieth century witnessed a move toward a theoretical and microscopic focus. For example, German physicist Max Planck (1858–1947) described matter as emitting discrete quantities of energy. He showed mathematically that matter was not fully separate from energy, a breakthrough in physics.

Newtonian physics discretely placed matter and energy as separate. Planck's ideas specifically called Newton's laws into question. Albert Einstein (1879–1955) elaborated upon Planck's descriptions, publishing the theory of relativity. In it, Einstein gave the mathematical relationship between matter and energy; this fully changed physics. Relativity described matter as being dependent upon more variables than

Figure 10.3. The theory of relativity: Matter curves space and time around itself. Source: Hemera/ThinkStock.

forces and objects. Time and space are variables that are relative to the object or person. A good way to explain relativity is in a person's perception of time. A first grader thinks the first grade took forever to complete, yet someone who is eighty years old sees the year's passage as rather quick—time's passage is relative to the person who is experiencing it. Einstein showed that these variables are relative to different circumstances. He showed, for example, that matter instead curves space and time around itself. Astronomers have confirmed Einstein's theories in their data on black holes and patterns of movement in outer space.

Einstein showed his ideas mathematically and not experimentally. There has been much confirmation by more recent scientists. Einstein also showed that matter contained large amounts of energy. A drive to find this energy within the atom led to atomic fission (splitting of the atom) and the discovery and development of radioactivity (energy emerging from the splitting of the atom). Scientists built upon Einstein's predictions to move theoretical physics into experimental physics.

While the positives and negatives can be debated, instrumentation, techniques, and discoveries emerged in the twentieth and twenty-first centuries from Einstein's postulates. To illustrate, the use of X-ray radiation led to the discovery of chemical shapes in three-dimensional forms. Body parts, from heart vessels to lungs, are imaged using X-rays and radioactive dyes.

The discovery of three-dimensional shapes led to some great results. The field of genetics emerged from the three-dimensional shape of DNA (deoxynbonucleic acid) because it showed the process by which proteins are made. James Watson and Francis Crick discovered DNA structure in the mid-twentieth century. These scientists used data to form a model of DNA.

DNA'S DISCOVERY: A CASE IN GENDER BIAS?

Much in the same way most areas of science were dominated by males, science in the discovery of DNA showed a similar pattern. Watson and Crick in 1953 used information from a variety of sources to discover a new model for DNA. While traveling to a colleague's lab, Watson discovered an X-ray image of DNA taken by Rosalind Franklin, who worked in that lab.

Watson studied the image to determine the shape of the double helix model. Franklin did not receive credit in the publication of the model. Unfortunately, Franklin died shortly afterward, receiving little benefit from her work.

DNA AND THE TWENTY-FIRST CENTURY

Their model of the DNA structure explains a great deal about how DNA divides and makes proteins in a cell. Within the past twenty-five years, the Human Genome Project sequenced twenty-two thousand human genes. Sequencing the genes is important because the code tells us what kinds of proteins are made that cause disease. If we know the proteins, then we can figure out the treatments against them. In 2012, the microbes on and within humans—our microbiome—were found to contain eight million genes. In fact, we have over ten thousand species of bacteria in the human microbiome of healthy people and between two and six pounds in weight of bacteria.

Watson and Crick showed that DNA looks like a twisted ladder with chemicals that match together making up the rungs of the ladder. This type of structure is known as a *double helix*. The "code of life" is contained within a string of base sequences that give rise to proteins. The model of DNA as sketched by Watson and Crick is shown in figure 10.4.

Figure 10.4.   Deoxyribonucleic acid structure and base pairing. Source: iStockPhoto/ ThinkStock.

BREAKTHROUGHS TO "MICRO" SCIENCE

Proteins made by DNA carry out all of the activities of the body and make up many structures. Some examples include hair, nails, enzymes to digest food, hormones to communicate internally, antibodies to defend against infections, hemoglobin, and gases in the blood. Molecular science is based on studying the proteins of living systems. If these could be understood, so many diseases could be cured. Progress in studies of the human microbiome could lead to cures in Crohn's disease, colitis, irritable bowel syndrome, and skin disorders. Understanding the genetics of human genes and microbes should yield cures for the future.

"Micro" studies of substructures in cells will help develop newer medicines to treat, among others, hormonal imbalances and blood disorders. For example, hemophilia is treated quite successfully by using genetically produced clotting factors to control bleeding. Human growth hormone and insulin are also produced by gene technology en masse to treat diseases on a larger and safer scale.

The shift to a micro approach to study smaller-scale science is an accelerating movement. Nanotechnology, which looks at structures one-millionth the size of normal, can yield big results. Large amounts of information can be obtained for each study using nanotechnology, for example, thousands of drug treatments for cancer could be tried in one shot. In a lab, a set of cells may be placed onto a test plate containing thousands of chips containing drugs. Many experiments are done simultaneously in what once were separate procedures.

Physicists are studying particles smaller than nano-level with smaller-than-atomic matter: quarks. Quarks, strings, and dark matter are also being studied to detect different levels of reality. Because it is believed that dark matter makes up a large part of the universe, it is possible that developments in these areas of physics will lead to major shifts in how we view the universe.

Science progress in each period of history or in each area of science is beyond the scope of this chapter. However, it is clear that modern globalization and the rapid advances in communication in the twenty-first century hastened science progress and collaboration among different groups. On the one hand, the decline in science literacy among the U.S. populace presents a new threat to science advancement. On the other hand, there is optimism in the rate of scientific advance seen so far in human history.

RAPID PROGRESS IN SCIENCE DISCOVERY

Truth always prevails, as shown in science history's chain of events. Science is driven by human curiosity, which could not be held back in any era. The success of truth was well expressed by Buddha in the statement, "Three things cannot be long hidden: the sun, the moon, and the truth."

Rapid progress within a short time period is science's miracle. The age of the universe is estimated to be fifteen billion years. In terms of geologic time, humans existed as a species on Earth at only the very end of this time scale: sixty to ninety thousand years. A single individual's life span is no more than a century in duration, if one is lucky. Yet a person can accomplish such a great deal in a short time. Aristotle, Plato, Newton, Galileo, and Einstein produced a wealth of philosophical and scientific changes in their lifetimes.

Carl Sagan (1934–1996), an American astronomer and astrophysicist, compared human time on the Earth with the long life span of the universe. If the beginning of time to the present is compressed into one calendar year, the following would be equal values: fifteen billion years = one year; one billion years = twenty-four days; one second = 475 years.

Only in the last hour-and-a-half of the last year would humans have entered into the picture. Human history since the ancient Egyptians would represent only a second of time in the year's life span of the universe. Even less, a human life, far fewer than 475 years, is but just a moment, less than a second. And so, in light of this, human accomplishments in science are so much more impressive. Some key dates follow in this comparison[8]:

Pre-December Dates
January 1—Big Bang (start of the universe)
May 1—Origin of the Milky Way galaxy
September 9—Origin of the solar system
September 14—Formation of the oldest rocks on the moon
September 25 (approximately)—Origin of life on earth
October 2—Formation of the oldest rocks on earth
October 9—Date of the oldest fossils (bacteria and blue-green algae)
November 12—Date of the oldest fossil photosynthetic plants
November 15—Eukaryotes (first cells with nuclei) begin to flourish

Late Evening, December 31
10:30 p.m.—First humans
11:00—Widespread use of stone tools
11:46—Domestication of fire by Peking man
11:56—Beginning of the most recent glacial period
11:58—Seafarers settle Australia
11:59—Extensive cave paintings
11:59:20—Invention of agriculture
11:59:35—Neolithic civilization, first cities
11:59:50—First dynasties in Sumer and Egypt, development of astronomy
11:59:51—Invention of the alphabet, Akkadian Empire
11:59:52—Hammurabic legal code in Babylon, Middle Kingdom in Egypt
11:59:53—Bronze metallurgy, Mycenaen culture, Trojan War, Olmec culture: invention of the compass
11:59:54—Iron metallurgy, first Assyrian empire, kingdom of Israel, founding of Carthage

11:59:55—Asokan in India, Ch'in dynasty in China, Periclean in Athens
11:59:56—Euclidean geometry, Archimedean physics, Ptolemaic astronomy, Roman Empire, birth of Christ
11:59:57—Zero and decimals invented in Indian arithmetic, fall of Rome, Muslim conquests
11:59:58—Mayan civilization, Sung dynasty in China, Byzantine Empire, Mongol invasion, Crusades
11:59:59—Renaissance in Europe, voyages of discovery, emergence of the experimental method in science

SCIENCE IS SEEN AS SOLUTIONS

Because of the dramatic effects of science on society (cars, trains, life-saving medicine), people have been lured into a sense that all things can be solved by science. To be sure, this past century witnessed an age of rapid scientific advancement. The ability of technology and science to better our lives and the lives of the next generations is very often relied upon.

Solutions to problems that have plagued people, from disease and earthquakes to outer space threats, are presumed to be within grasp. This optimism is not without good reason. Science has offered solutions for so many maladies. We rarely hear about smallpox or the measles, but many died only one hundred years ago from these maladies. In fact, infant mortality is almost one hundred times less frequent, and life expectancy has almost doubled (forty-eight to seventy-eight years) since 1900.

America's drive for buying technology and plastic surgery and the demand for better, greener energy exemplify our culture's excitement with what science has to offer. As far back as 1893, biologist T. H. Huxley (1825–1895) expressed the new age of scientific confidence, stating,

> I see no limit to the extent to which human intelligence and will, guided by sound principles of investigation, and organized in common effort, may modify the conditions of existence . . . and much may be done to change the nature of man himself.[9]

Scientists verified Huxley's optimism with a continuous stream of products and services. Research advanced in many fields. Discoveries in chemistry and physics laid the groundwork for future advances.

RAPIDLY ADVANCING SCIENCE
IN THE EARLY TWENTIETH CENTURY

Anyone one hundred years old today witnessed rapid scientific developments. Between 1900 and 1950, automobiles, telephones, television, airplanes, electricity, rockets, atomic bombs, radiation, jet propulsion, robots, computers, radio, and even Silly Putty were developed and put into widespread use throughout the world. The first half of the twentieth century was the most productive set of years than any other in human history. Many scientists today are envious of the creativity and genius that drove scientific change in that era, and its significance is often underreported.

Life changed from having horses and no running water or electricity to the modern society of rapid transportation and communication. Smartphones and the iPad are nothing in comparison to the changes back in 1913. People could not travel by car or jet plane before this time. Energy usage changed from wood to petroleum products. Alternative energy, such as hydropower and nuclear fuel, was also developed during this time.

DANGERS

Not all of science's results are useful or even good for the public. Weaponry became more dangerous, with rocketry and nuclear threats of radiation as a new fear. This ushered in an age of fear and doubt about science along with the age of optimism. It became obvious that science had perhaps sown the seeds of human destruction.

The more people used science, the less sure science made the world. Even Einstein's theory of relativity stated that time and space were relative, not fixed or certain. Thus, even motion and reality cannot be simply understood: Is there another universe, as Einstein posited? Could humans travel back and forth through time? A great deal of science fiction and wonder grew in society.

DECELERATION

Today there is a deceleration of scientific excitement and progress. In history, the creativity and invention of a small handful of scientists led to tremendous change. Almost everyone has benefited from twentieth-century advancements. However, most people are merely operators of technology and do not understand it.

They know how to use TV and computers but do not know how they really work. It is perhaps this passive acceptance of science that has led to its own deceleration. Science progressed rapidly in specific areas during the latter half of the twentieth century. At the same time, science literacy among the population declined.[10]

Since 1950, science advancement developed from old technology that was merely improved upon. For example, the *Apollo* space missions to the moon in the 1960s and 1970s relied on rocketry. But rocketry was developed long before, during World War II in the V1 and V2 styles. Medical breakthroughs use radiation, echo waves, and imaging that date back to developments in sonar and atomic weaponry of pre-1950. Computers were around before World War II, albeit larger and less sophisticated.

Automobiles are pre-1950s technology—it's shocking that a 1950s automobile was simpler and, in many ways, easier to fix than our modern cars. Autos are safer and more efficient (some models) than in the past, but they are still an old innovation.

As a child growing up in the 1970s, I watched a cartoon about a future world called *The Jetsons*. It had rapid, casual space travel, time travel, and teleporting, with robots to do people's work. The cartoon expressed optimism for science, which was the sentiment of the public in the mid-twentieth century. Obviously none of these developments materialized. While it is immature to expect time travel and teleporting, who is to say that it could not be developed? What is necessary is the recruitment of great minds to drive science.

There were certainly scientific advances made after 1950: the flight to the moon, medical breakthroughs, and computer improvements. However, these were all improvements of prior technologies and knowledge and not the results of great new insights. Deceleration is a hypothesis poorly received by modern scientists because it somehow insults them and their research community. This is stated not to minimize their accomplishments but to highlight the changes in culture that have led to an overall slowing of creative genius in science. After 1950, the rate of new discovery and innovation declined sharply, and its effects are of international concern. Science deceleration is carefully studied in this book to prevent its threat to science progress. There are ways to recruit excellence into science, which will be discussed in the next chapter. This is a key factor in reversing the recent trend.

LESS OPTIMISM

The results of science deceleration are alarming: automobiles are bigger than ever and there is more gas guzzling as the SUV has taken the place of regular cars. Transportation is accomplished through the same methods as in 1950: cars, trucks, roads, trains, and airplanes. Yet in the past ten years, travel by cars and trucks increased by 17 percent in the United States—a 10 percent increase in the number of drivers, a 35 percent increase in the average number of miles driven—but federal and state governments have constructed only 1 percent more roads.[11]

We do not have mini rocket ships or teleporting for rapid movement. Nuclear warheads are pointed in every population center in the United States, and no missile defense shield was ever effectively developed. Technology is an obsession, replacing the underpinnings of science interest. Users of technology fawn over cell phones and computers while science understanding declines. In fact, curiosity about the natural world is in decline as evidenced by the United States' recently giving up its NASA space shuttle program and any plans for a return to the moon. The lack of excitement about science is illustrated by this move. The space program reflects the nation's commitment to discovery, and that is the driving force behind science.

The case for a deceleration in innovation and scientific creativity underscores the importance of a continued focus on science as a solution. There are many threats to science: one major roadblock to its progress is an elevation of the nonintellectual. This will be discussed in the next chapter. While economic progress in the next decade will depend on the next scientific innovation, the literacy of science will define the quality of our culture. Learning and creativity need to be supported in the pillars of our society—education, family, religion, and government—to develop new Einsteins and Galileos. This is the road to scientific progress.

THE NEXT BIG INNOVATION

It is impossible to predict the next big innovation that will create a paradigm shift within our culture. Future scientific developments cannot yet have a name. They are not taught and have no impact on our economy because they have not yet been invented. They are still in the minds of the creative genius of future generations. Perhaps they are in the neuron workings of babies being born at this very second. It is impossible to know what science will be like in the future. How science developed in the past gives us a clue about the innovations that are yet to come. We know from science history that scientific innovation always drives economic progress and military success: rocketry, the printing press, gene therapy, and microchips.

The next big science innovation depends upon the state of our culture. If present conditions remain the same, energy science will drive the future. It comprises a large part of our economy now. A small powerhouse-style box to give electricity to each house would have dramatic effects on our lives and the economy. Energy would become cheaper, and resources would be more easily obtained. Our standard of living would rise.

If society remains the same, information technology will continue to help the economy on a number of levels: inventory management, shipping, marketing, and communication. In addition, the Internet allows for a wide range of consumer uses: shopping, gaming, communicating/socializing, organizing/storing information, and locating information. These developments will not drive the future in the long term.

Conditions and variables in a society never remain the same. Change is as certain as time's passage. It is futile to make specific predictions about future innovation with much accuracy. General areas in which we might see major developments are robotics (consumer, commercial, and military applications), nanotubes, commercial rockets/orbiters, biotechnology, cloning, artificial intelligence, and new weapons for the military.

These extrapolations are taken only from our view of science and society as they are now. If a massive outbreak of a mutated virus kills 90 percent of the population, I doubt energy research will be at the forefront. Societal change will determine science's progress, as shown throughout its history and in this chapter.

However, there could be some discovery, by chance, that changes the course of science and society. Could it be nanotubes? Nanotubes are composed of flexible carbon sheets. They may be used to construct lasers that zap tumors, build hydrogen fuel cells for cars, and perhaps form a long (sixty-two thousand miles) space elevator to the moon. Chemical properties of the nanotube are remarkable, with flexibility and a nanostructure to get anywhere we wish to place it. Could it be termite gut bacteria? These microbes are able to digest lignocellulose, which would yield more efficiently much of the energy from wood products. More specifically, it would improve ethanol production from corn to make it more profitable. Could it be a discovery of a twin of the Earth? Thus far, over five hundred planets outside of our solar system have been identified as possible places for life as we know it. It depends upon their distances from the stars they orbit, as well as their supposed atmosphere and composition. Life is water- and carbon-based, so the new planet would need these prerequisites. One such planet, Kepler-22b, has been identified but would take several hundred thousand years to reach by current rocketry. Alien DNA? A recent discovery of DNA in Mono Lake, California, indicated that an arsenic-based DNA structure was found, perhaps belonging to aliens. Of course, it was dismissed as a contaminated substance. But such possibilities of alien life would change the course of science and our civilization.

What can be done to encourage the next great revolution? Who will drive it and why? Possibilities are endless, but they begin with attracting new talent into science. Tax breaks, tuition assistance, research and development focus, or changing cultural norms—central to these brainstorm ideas is the development of a scientifically literate populace. Human capital is by far the most important asset to any society's science.

There is a critical number of people needed in science to make these changes. No governmental policy will be more important in furthering science than one that encourages a change in culture to attract new scientists.

11

Getting People to Love Science

Choose a job you love, and you will never have to work a day in your life.

—Confucius

The wise words of Confucius remind us that science needs to be enjoyable to people in order for them to choose it as a profession. Recruiting and retaining smart, effective contributors is what will drive science for the next generations. Science has to be "sold" as cool for people to want to do it. This chapter is dedicated to evaluating the structure of our culture to determine the elements of it that may turn on and turn off potential contributors to science. The last chapter focused on the leaps and potentials science has to offer our society. It looked at science's past of great scientists and remarkable progress in a relatively short time. People must study and do science to make that progress possible. Only through moving toward a science-oriented culture will more human capital be directed toward these goals.

What are some roadblocks to attracting people into the sciences? All of the recruitment problems in science can be traced back to one key element—a lack of support for future scientists. Here are some reasons STEM (science, technology engineering, and math) appeal is flagging in the United States: 1) the media-educational complex teaches people to make fun of scientists; 2) the media-educational complex places higher value on sports and entertainment as a diversion from academics; 3) heroes in science are dwarfed by Hollywood and sports stars; 4) an ethos valuing science and intellectualism declined in the dominant culture over the past century; 5) technology's misuse guides toward less learning and more entertainment; 6) values of hard work (which is what is required to be successful in science) declined over the past century; 7) the work environment for science jobs is less pleasant than in nonscience; 8) critical-thinking and science literacy skills are weak and make science inaccessible to most of our public; and 9) science teaching does not cut the mustard to attract and retain students.

A focus of this chapter is to analyze the many threats to modern science and then to recommend ways to improve upon recruiting and retaining future scientists. Throughout history, scientific thinking has always attracted controversy. Real science is apolitical and "outside-of-the-box" in its thinking, but politics and power interests pull science into the fray.

Science also often overturns the existing structure of thought and society through its discoveries. For example, the development of the automobile changed society; the World Health Organization reports that 1.2 million people die worldwide every year as a result of automobile accidents—more deaths than result from many diseases. Yet people are still willing to use this technology to improve their life functions. Science radically changed society through automobile development, regardless of the negatives. People use technology and science despite its dangers, calculating risks of benefits versus costs of use. Science has controversial impacts on society which enhance conflicts with its effects.

Science and its products, despite drawbacks, are intrinsically valued by our society. We beg for the latest medical procedure if we need it and we use the latest computer application to improve our business's bottom line. Then why is science so poorly regarded that it is hard to recruit enough new talent to fill the void? Let's explore the roadblocks to attracting people into the sciences.

ROADBLOCKS TO STEM

There are three general elements which contribute in influencing people about science areas: our culture, science as a profession, and science education. Each of these aspects has features which both appeal to and turn off potential recruits. It takes an integrated analysis of the interaction between these three branches of society to understand how STEM success occurs within a culture. The individual does not exist in isolation; many factors influence a decision on career choice. The three factors—culture, career, and learning environment—play a central role in identity development. From childhood, these three interact to persuade interests in many different areas. Children learn best by example, with an emulating mechanism to follow the interest of adults and other children surrounding them. Let's give a look first to the cultural components that influence people's opinions about STEM areas and careers.

Roadblocks to STEM careers did not develop overnight. In this section I present the changes in society over the past century contributing to declines in STEM achievement and deceleration in science. Particular attention is paid to declines in science achievement in the United States since the 1950s. Concomitant with these declines are a set of features in the larger culture that weaken interest in STEM as a career and a driving force for the economy.

I. CULTURE

Science Images

Culture plays such a vast and complex role in shaping opinions, attitudes, and interests. Examination of the media, educational institutions, and government roles and politics helps to explain the changes in STEM interest. The way in which certain careers are depicted through these elements of society gives young minds their perspectives. How are STEM-related careers portrayed and valued by the dominant culture?

In our larger society, there is an abundance of socially unattractive images of science people. For example, students, scientists, science teachers and scholars are shown by media outlets as strange and unappealing. Very often, science professionals are shown with horrible faces, crazy unkempt hair, or unattractive features and eccentric demeanors. They are set up as characters to be mocked but certainly not to be admired. According to several studies by NSTA (National Science Teachers Association), the images children have of scientists affect their attitudes toward science. Negative images of scientists were prevalent in a study asking students to draw a scientist. These images came from media depictions because most of the young people in the studies had never met a science professional.[1] A cultural transmission of negative stereotypes was detected, which is fomented by the media-educational complex.

Consider the sitcom *The Big Bang Theory*. Science characters are shown as misfits, relegated to a secondary social class status. While surficially entertaining, the program portends the future of students thinking of entering a science field as one of alienation and ridicule. Potential recruits to science need positive images of successful science professionals who are respected within the larger community. The draw of Hollywood or sports characters emerges from the media-educational complex, which constantly depicts celebrity status as something to be desired. Great effort is put forth by this setup to gain popular support for celebrity dreams but little or negative support for science.

The Big Bang Theory sitcom is not the problem; in fact, it can be humorous. At least science is being depicted within our TV curriculum and there is some attempt at showing some positive antics of the characters. However, the underlying appeal is that the program, much like other depictions of science in society, castigates intellectuals. Being "smart" is something to be ashamed of instead of something to be proud of. There are numerous studies showing this tendency in our society. It is apparent whenever popular media programming reveals its emphasis on anti-intellectualism: a majority of programming is entertainment centering on sports and lower level reasoning.

In contrast, many other cultures venerate STEM achievement. Scores in international mathematics and science tests, as discussed in chapter 1, "Introduction," show other nations far exceeding U.S. achievement in STEM areas. Unfortunately, as will be discussed in the next chapter, it is to our economic dismay that such a trend is tolerated by the larger culture.

Anti-intellectualism and Ageism

What are some cultural reasons for a decline in STEM images? First, it is based on an underlying anti-intellectual sentiment that pervades our society. Wisdom and experience are not as valued as in other cultures. In Japan, for example, the elderly are given a great deal of respect because of the wisdom they naturally gained through years on this Earth. Instead, in the United States, ageism is tolerated in ways that other forms of discrimination would not be. Recently, a sixty-eight-year-old woman was bullied and tormented by a group of middle school students on a school bus in Rochester, New York. There was an outpouring of public support in 2012 for the lady after the "children" placed a video of the bullying on YouTube to brag about their actions. This shows how ageism is rampant but that many of us find it intolerable.

Ageism is a virulent part of American society, as seen in the elevation of youth by the media. The drive to stay young and remain a part of the social order has led to a boom in cosmetic surgery and shifts in the fashion industry to propel a youthful image. Most TV programs are geared toward eighteen- to twenty-eight-year-olds, with older characters rarely in dominant roles. Marketing products also are geared toward young adults, who are shown to spend more than other age groups. Thus, a drive to capture this market share results in a widespread media culture geared toward the young. Money may be driving the media-educational complex to push youth, but the result is destructive for science and science education recruitment.

With youth comes a certain lack of wisdom and understanding. This is not to say that there are not brilliant and wise young people. However, age and experience correlate with competence and skill development. On the other side, electroencephalography shows that certain portions of the brain controlling aggression and risk taking are completely underdeveloped in the adolescent brain. Thus, young should learn from old so that the wisdom of the ages is passed along. Instead, the gap between the generations increased markedly over the century. While on the old homesteads and in immigrant families, multiple generations lived together. Young people learned from their grandparents and a kind of intellectual sharing took place.

In today's fragmented society, the generations are taught not to speak; that somehow they cannot relate to each other. Grandparents rarely live with their grandchildren, in contrast to the past. Many move into communities or elder care facilities restricted to seniors, which exacerbates this disconnect between the generations. The idea was fomented that young and old do not mix and have nothing in common, that somehow a person is not normal if they relate to or listen to older generations.

The decline in intellectualism, as a result of the disconnectedness between older generations and the young is obvious. Older generations have a great deal of wisdom to share with younger generations. This wisdom is a necessary precursor to creating mature prescience professionals. Maturity is necessary to successfully engage in science education, which is arguably more rigorous than many areas of study.

The diminished connection with older generations resulted in a lessened transmission of some of their values, which are needed to make good science students and scientists. Attitude is more important than intellect and, as Albert Einstein one stated, 90 percent of success is gained through effort. In the past, teachers, professors, scientists, scholars, and intellectuals were more respected by prior generations. The key to STEM entrance and retention in a professional field is having the kind of respect for the authority figures (teachers, bosses, parents, professionals, etc.) that was prevalent in generations of the past.

Programs to connect older and younger participants vastly help in bridging the gap between the generations. Young volunteers in elder care facilities have been shown to benefit both groups. For the young, this augments the kinds of skills necessary for STEM students to be successful in their studies. To be successful in a STEM area, it requires serious academic pursuit and maturity. It requires that a student be serious and intellectual, aspiring to a higher level of learning. Because wisdom and intelligence is often ridiculed by the media-driven culture, the attributes of a good scientist are also ridiculed.

If the larger culture does not value what STEM areas require—intelligence and focus, which is obtained by interacting with elders—then an expected result is a decline in people prepared in those areas. Until the culture changes its focus from negative to positive images of science, people will shy away from science. Similarly, images that are given a favorable depiction lure students into their fields. We should emulate the advertisement that sports and Hollywood use to imprint upon science promotion.

Competing Images with Science

What areas are shown positively in the media? Nonscience areas, on the whole, are shown as cooler and as more interesting than STEM subjects. Most alluring, of course, are sports and acting careers, which show successful participants as "stars." There is nothing wrong with these careers, and both are valuable to our society. It is deceptive to bring so many people into areas which have low demand and low income-earning potential on average for the field, yet these careers are sold as the ticket to wealth and fame.

STEM jobs have much higher salaries than non-STEM jobs. Rising student debt and wastes in time and energy for fields with low demand require that school strategies for recruitment into STEM areas change. In chapter 12, "Driving the Economy through Science," economic gains for a person entering a high-demand STEM area are explored. The first step is to convince colleges and universities that schools should produce more STEM majors to fill the nation's needs. Giving realistic expectations upfront to students entering low-demand professions should be an obligation in all levels of instruction.

Instead, there is an attitude in education that STEM areas are too expensive and cut into the college's bottom line. The president of the University of Florida at Gainesville recently announced that STEM majors should pay more tuition to offset the higher costs of laboratory fees and instrumentation associated with their education. Would higher education rather lure students into fields that will leave them saddled with debt and with less job potential, instead of recruiting more students into the STEM majors?

Whatever field is respected by the culture will draw recruits and produce graduates. Who is respected by the dominant culture? It is clear that it is not science educators or scientists that garner the high respect. Two groups of job fields have favor with the public, usually those who are deemed to be in power positions: 1) the "power group": military, police, firefighting, and health, and 2) those fields which are glamorized, termed the "fame- and-fortune group": sports, acting, and even politics. Each of these professional groupings is venerated while science is depicted by the media as "nerdy" and "crazy."

These respected fields with favor should not be torn down in this chapter. There is a place for each of these professions in our society. It is not the respected professions that detract from science but rather the media-educational complex's emphasis that certain professional fields are valued over others. Favoritism of this sort leads the public to yearn for social acceptance and entrance into those areas. When a person says she wants to become a doctor, for example, she gains more respect than if she expressed an interest in studying digestive physiology in the lab. It is in part, this need for "good PR" (public relations) that ruins basic science interest in our culture.

The "power group" is continuously depicted as the people in charge. They come in and save the day. Crime dramas abound on TV (*NCIS, CSI, Law & Order,* and *Blue Bloods*) and law enforcement characters are shown as in control and being given social respect. They are well integrated in society and have attractive physical features. In each episode, the characters wear stylish clothes, attract many friends, and have money and prestige.

This brings us to the allied health professions. These fields are important STEM areas and interest in them has increased greatly in the past decades. They buck the media assault on science. Allied health professionals are shown as smart and in control. A doctor is still given great respect by the dominant culture, student interest in these areas continues to grow. In part, it is because salary is high for these fields and people always respect money.

According to the Higher Education Research Institute at the University of California, Los Angeles (UCLA), student interest in majoring in allied health fields doubled since the late 1990s. The number of students interested in nursing is at a twenty-eight-year high (3.9 percent), and at all-time highs in pharmacy (2.4 percent) and dentistry (1.1 percent).[2] All allied health fields come with higher annual salaries and better demand for their skills; thus, they are still coveted and respected. Medical school applications rose dramatically and physician's assistant programs are filled to capacity.

Harnessing this energy to drive people into the sciences should be a major goal of educational systems and the government. Their advertisement is a template for which STEM image changes should be made. Perhaps incentives to move some clinicians into basic research and development would be valuable in forwarding science discovery. What is it about the allied health field that prevents it from being denigrated like other science areas? It usually comes down to money.

Economic Attraction to STEM Careers

The "power group" and the "fame-and-fortune group" offer large amounts of money, in the form of salaries, to enter their fields. This is a main reason so many people want to go into those areas. This is also why they are low-demand fields, leaving many without entrance into them. This lack of access to low-demand fields should be better explained to those beginning to choose a career.

Also, in order to improve the image of STEM careers, salaries should rise to meet their demand. If there is a shortage of STEM teachers, researchers, and health professionals available to meet the nation's needs, then salaries will rise with supply-and-demand principles. Teacher pay is still lagging in STEM areas and shortages have been compensated for by allowing poorly qualified teachers to teach those subjects. For example, recent data show that roughly 40 percent of teachers in science areas do not have a major or minor in the field in which they are teaching. Almost 60 percent of high school physical science classes are taught by out-of-field teachers. Within STEM areas as a whole, 47 percent of secondary students were taught by teachers without a major or minor in their teaching assignment.

Variances and waivers are granted to these instructors to simply fill the void of qualified teachers. Is a person who merely took chemistry qualified to teach it as a professional? Is this a good enough role model in science for our kids? Because this is the case in one-third to one-half of our science classrooms, standards for teachers ought to align with science education reform.

Due to a lack of respect and money, the teaching field is attracting students with worse academic abilities on the whole. Education students trail in their average SAT scores by 49 points, with a mean SAT in verbal and math of 850. The best and brightest are not being lured into STEM teaching.

Poor teacher preparation is not shocking when non-STEM teachers and coaches are paid the same or more than STEM teachers. Why work so much harder, taking laboratory courses and mastering difficult material, only to be unrewarded financially as well as socially? The educational system is set up as a socialist, heavily unionized shop, with all members garnering roughly the same wages. This is an intrinsic disincentive for STEM instructors, who could make much more money in the capitalistic private industry that values STEM skills.

Most people respect money. If a football player makes millions of dollars a year, people will brag about him and want to pay to see him play. Because money is a driving force in respect development, STEM teachers need to earn more money than non-STEM teachers. By allowing the laws of supply and demand to take hold, the quality of STEM educators at the secondary level will increase dramatically. This is not said to insult non-STEM teachers. They are on the same side of intellectualism, which fosters science achievement. It is stated because STEM teaching needs recruitment from alternate means of employment in the industry. If pharmaceutical, medical, and engineering positions are high-paying, STEM teaching salaries also need to be appealing to draw talent.

As a result, STEM teachers, statistically, are often not properly qualified, and this impacts secondary preparation vital for success in college science. It is a partial result of respect, which, for STEM teachers, is particularly lacking in the dominant culture. Wages reflecting the high demand for STEM teachers would raise their level of respect.

Consider that college coaches and professional athletes earn, on average, five to ten times the salary of a college professor. If this differential is tolerated, the society will get what it deserves: continued decline in U.S. competitiveness in science and mathematics and a secondary role in the world.

Allied Health is Still Respected: Links to STEM

The dominant culture values certain attributes of a career: respect, money, and prestige. As stated earlier, they all go hand-in-hand. Medical doctors get all three of these rewards in their professions, albeit the education and commitment is extensive and the rewards are well-deserved. The life of allied health professionals is tough, with a great deal of work, time commitment, and stress. However, the images of health professionals remain positive in the dominant culture, overall.

Medical doctors save lives and intrinsic in this fact is a certain amount of respect and reward. Unfortunately, the sciences are often seen merely as a conduit to the allied health fields. Biology education gains most of its members through hopes for a career in an allied health field. Rarely do I have a student in my anatomy and physiology course who wants to simply take the class for personal fulfillment or for a future in medical or science research.

The problem is not that recruitment into allied health is bad; this recruitment does, however, provide evidence of the power of culture in dictating student initial interest and retention into a major and a profession. However, allied health remains allied to all science because, at the end of the day, doctors are scientists. This link should continue to be emphasized to improve the image that scientists have in the media and culture.

If society was really serious about improving STEM participation in the work-force, it would make images of these areas more attractive by enhancing the payoff for entrance. If a scientist earns more money than a football player (millions/year) or even a policeman (in some areas over $100,000/year), then the culture would follow in respecting the scientist. More qualified people would enter the field and STEM would be a more effective force in driving the nation's economy. President Barack Obama's goal to have 100,000 more STEM jobs through Educate to Innovate would materialize. For every non-STEM job, there are 3.6 unemployed people whereas STEM jobs have double that demand. Right now, too many students are preparing for jobs in which there is low demand.

Accomplishing this improvement requires a marked shift in cultural promotion of careers: schools should push science competitions and science fairs as stylish; TV should have more science shows noting discovery with attractive characters, not just medical dramas; government advertisements should portray images of science people as successful; industry should do its part through advertisement linking its science careers to success; and the media must draw positive images of elders, presenting intelligence and science as respected. The best way to accomplish this is by redistributing where money is spent. Marketing efforts by power holders in society should redirect money to science-related efforts and away from low-demand fields. In this way, the public will follow the trend set and STEM education would garner more respect and attract more talent.

Multiple Intelligences

The second classification of job choice that is respected in the culture is the "fame-and-fortune group": sports stars, Hollywood stars, and affiliated careers (e.g., news broadcasters, sports writers, announcers, coaches, singers). All are well-paid but extremely low-demand jobs. Everyone wants to be a movie star, but few will attain this goal.

Very few people who try for a sports or Hollywood career are actually successful (less than 0.01 percent or 1 in 10,000). In spite of this, the media-educational complex encourages people to follow their dreams and become whatever they would like. This sounds very curmudgeonly, but the media-educational complex is luring people into disappointment. It is creating the student goals in the first place and then duping them. By depicting "power group" jobs and "fame-and-fortune group" jobs much more positively than those in STEM fields, it is forcing these goals onto public focus. Young minds are impressionable and media-educational techniques create a culture of risk takers who would rather spend their time going for a long shot.

According to the multiple intelligences theory developed by educational psychologist, Howard Gardner, traditional intelligence is questionable. Instead, the ability of a person may lie within many spheres of abilities. We all possess certain skill strengths, some which are developed and some which require more effort to bring forth. Gardner proposes eight subsets of intelligences: spatial, linguistic, logical-mathematical, bodily kinesthetic (psychomotor), musical, interpersonal, intrapersonal, and naturalistic. Spatial ability allows a person to see three-dimensional conformations; linguistic or language skills involve the memory of words and the use of these words to create meaning; logical-mathematical talent is the traditional scientist's gift with numbers and patterns; psychomotor ability involves the use of one's body to accomplish tasks; musical ability creates sound to interact with other senses; and interpersonal ability facilitates human interactions whereas intrapersonal skill includes understanding oneself. Naturalistic skill creates oneness with nature and the environment.

Each of these natural intelligences enables a person to become adept within certain professions. Someone who excels in psychomotor skills, for example, would work well with one's hands or body. She or he may become a dancer or an excellent dentist or waiter. A farmer should have naturalistic ability but also needs psychomotor skills to farm and logical-mathematical skills to conduct the business of a farm. A writer writes, so linguistic talent is required, but also a logical-mathematical talent if the writing is about science.

Obviously these are quite diverse recommendations for career choices. However, it is a confluence of different aspects of these intelligences that creates a person's predisposition to a career. No one person is only skilled in one area. Thus, multiple intelligences make up a person and a personality. A recent tight rope walk by Nik Wallenda, for the first time in history across Niagara Falls, showed tremendous bodily kinesthetic ability. A fear of heights may prevent someone else from such a feat. Someone with similar bodily kinesthetic talent, ability, and skill but with a fear of heights might instead find success as a cyclist.

A satisfaction with developing one's talents is important in the drive to recruit people into STEM areas. Many people have intelligences in areas that could be useful for STEM professions. A person with great interpersonal and logical-mathematical skills might be a candidate for a career in neurophysiology or brain functions or as a popularizer of STEM ideas and products. Clearly there is a logical-mathematical foundation underlying STEM areas, but talent in supporting intelligences is what is needed for the many STEM-related professions. The professions of psychiatrist, physiatrist, physicist, agricultural researcher, and science educator all draw intelligences from many of the areas mentioned.

There are cases in which logical-mathematical ability is so developed among science people that other areas are lacking. An ability to work interpersonally is vital for teamwork in many industries. STEM areas seek talented people who have skills in more than just one area of logical-mathematical ability. A successful person in a STEM area is often able to draw from multiple intelligences to work through science problems or gain knowledge and discover.

It is cliché to think that STEM is all about science and mathematics. How often have we seen a real person support the science stereotype image? A person with only logical-mathematical skill is not well suited for teaching, which requires interpersonal and intrapersonal skills for working with students. It is important to develop the liberal arts major to attend to the many skills within a person. Even if someone in STEM is suited heavily suited toward a logical-mathematical disposition, the liberal arts education can help break that person out. By developing other intelligence areas, that person becomes a better contributor to STEM fields. How boring to think that all science should be mathematical-logical and lack individuals with interpersonal or artistic talents. This is why STEM education and STEM employment should attract and utilize the multiple skills people possess.

The goals for the public should be to view STEM not as out of reach to the common person, but within us. Everyone has talent in areas that could contribute to STEM fields. Those intelligences need to be developed in each of us. Science education should have, as a central standard for instruction, an ability to bring out the talents necessary for success in a STEM field that are within each of us. Once there is public acceptance that STEM is within reach of all people, STEM will become more integrated into our culture.

This will prevent the long-shot goals that many rising students possess. If STEM fields, which are solid-paying and inherently interesting, are within reach of all people in some capacity, then students will shoot for those goals. However, it comes down to the way one views oneself. If students believe in their innate abilities, then they will garner the motivation to develop those abilities. It is the role of science education to help students view STEM as accessible and doable by everyone.

While the skill set to become a scientist is not in everyone, everyone is capable of doing science in their everyday lives. Analyzing grocery bills or evaluating food products require these same set of intelligences, though at different levels than STEM careers may require. Getting the public to love science because its skills are within them should be the goal of science education for everyone. STEM skills have a great deal to offer the public.

A cultural shift accepting STEM as valued and part of the dominant ethos will follow. If adolescents value themselves and believe in the multiple intelligences that they have, they will strive to develop those talents. It is a lack of confidence in adolescents that drives much of their behavior. Desire for social acceptance and justification drives their behaviors, from drug use to fashion trends. Through building up the individual to believe in his or her own intelligences, talent will be better able to grow. It is this set of intelligences that can be harnessed by STEM education to draw and retain quality students.

Competition and Science

Cultural drives to be the best and to work toward a long shot are in opposition to scientific advancement. We all have innate skills, as shown by the multiple intelligence theory. These skills include collaboration and communication. Fostering competition instead of collaboration leads to isolation. People working together instead of against each other best emulate the scientific ideal.

Recently, the media has pushed a competitive stimulus onto society. In the past, shows would have music with people singing and dancing together: *Sing Along with Mitch*, *The Lawrence Welk Show*, and *Dinah Shore*, for example. These fostered collaboration and dancing with couples. People demonstrated their psychomotor and musical talents together. Juxtaposing these images is today's *Dancing with the Stars*, *So You Think You Can Dance?* and *American Idol*, which adopt a competitive theme. The music is not about the beauty and skill of musicians and dancers but rather promotes a nasty survival of a subjective "best."

This works against the kinds of attributes necessary for success in a STEM career and a STEM education. Competitiveness is often cited as a reason students leave STEM education. While the industry is laden with competitive goals, the process of science requires building upon prior knowledge and working together with other scientists to develop great innovations. The Manhattan Project was a competition with axis powers for the nuclear bomb, but scientists on either side of the Atlantic worked closely together within their cohort. U.S. scientists and German scientists were each victims of espionage, but while each group was secretive, they used any information they could from the other side.

The media-educational complex needs to support a collaborative structure and not undermine it with reality TV and the hostile competitiveness in fashion. A society with solidarity around the importance of STEM areas will also lead to more practical goals by the public. The shift from "power" and "fame-and-fortune" careers to STEM careers will happen . . . but it will take commitment from the larger culture. Based on the status quo, we will continue along a trajectory which will place the United States in a weakened position. However, conditions and variables always change, geopolitically and economically, meaning that I predict a cultural shift to bring out STEM successes.

It could be a major discovery by another nation, it could be one of our own discoveries, or it could be a need for some important product to save lives—regardless, an impetus will arise to accelerate STEM achievement by our society or another. STEM is based on curiosity and the creation of new ideas and products. Curiosity and creativity are innate desires in humans. Power and control give glory to some career choices right now. But things change, and human nature's curiosity always prevails.

Self-Confidence and STEM

Our nature also asks us to be satisfied with realistic expectations. STEM careers are attainable, and this should be communicated to the public and prospective students. STEM education needs to become accessible through the changes described in this chapter. In these ways, more people will choose a STEM track instead of the disappointing low-demand careers of "power" and "fame-and-fortune." Improving STEM recruitment and retention will lead to economic recovery for the nation and the global community.

An old adage, *zufriedenheit macht gluecklichkeit*, translated as *satisfaction leads to happiness*, should be infused into the dominant culture. When people are satisfied and accept the talents that they possess, they will find suitable careers. They can only accept these talents if they realize their strengths within—that they are capable of developing those abilities. It is then that more people will have the confidence to do a STEM track and be successful in a STEM career. STEM careers are intimidating. A medical doctor's life or a professor's education both seems insurmountable. However, if believing in oneself is the key to developing one's abilities, then satisfaction with one's life is the key to happiness.

A lack of self-confidence to follow a STEM career has resulted in students following job paths that lead to financial ruin. Many think they are "no good in science" and brainwash themselves in a spiral of bad grades and embarrassment in science classes. It is, in many ways, a psychological barrier to success in STEM. STEM education is in part to blame, but the dominant culture plays its role as well.

Low self-confidence is responsible for the growing student debt and under- and unemployment. STEM careers would enable people to pay off their student loans and contribute to a successful economic life. Instead, time and energy is wasted on sports and non-STEM careers in the vital time of one's youth. American adolescents spend, on average, 7.7 hours per week in a high school sport and 4.9 hours in a nonschool-related sport. This time could be more productively spent in an academic pursuit. Over 82 percent of young adults participate in a sport in high school.[3]

While sports can be healthy and improve fitness (and sometimes self-esteem), they require time spent away from family and away from academics. I advocate sports and entertainment as an excellent diversion from academics and life. Hobbies, interests and physical activity are what make life interesting and worth living.

However, through an extreme emphasis on those areas, people miss out on an education and possibly a career in a STEM area. In contrast, imagine family spending time learning and growing together. Entertaining ourselves to constitute large portions of our lives is a shame. This trend has increased a shift in focus from academics to entertainment. As a result, science loses good talent and the people lose out on valuable interactions and life benefits that could be provided by STEM interest.

These patterns are tolerated and encouraged by the media-educational complex. The media portrays "fame-and-fortune" jobs positively while either ignoring or making fun of intellectuals. This creates a society that accepts the emphasis on sports and entertainment over education. In science education, it is particularly disturbing because so much of science builds upon the levels of education in earlier grades. Mathematics literacy is a prerequisite to science literacy. Without the work put into both areas, science education at higher levels becomes inaccessible to the majority.

Entertainment should be further considered because it plays a major role in shifting emphasis from academics to lost time and competition. Entertainment time has serious, negative consequences on our society and on the scientific recruitment and retention of good students. It is true that entertainment is a vehicle "to keep kids off the streets" and in the schools, with at least some rubbing off of academics in the process. However, this is not good enough. When international test scores show American children and adolescents falling farther and farther behind in mathematics and science, covariation with the entertainment trend is unarguable. The link between time and focus and academic achievement is well-documented. When entertainment detracts from academic study time (and it likely does, at 52 hours of sports/entertainment per week, on average, for a high school student, not including travel time!) then action should be taken.

While I commend any sports team that emphasizes academics, the problem is that the focus is still always on the sport. Academic pursuit is a means to an end; and the sport becomes more important to the player than the area of study. This phenomenon is peculiar to the United States, which includes sports in its secondary schools. In most of Western Europe, sports and schools are separate entities. Detracting emphasis from STEM education and into other fields is based on people's low self-confidence in science and academics. Empowering people to do well in STEM areas should be a central goal for science education in the twenty-first century. This would be the best way to combat this unfortunate but rather modern trend.

One Hundred Years Ago

While we have made great progress from over a century ago, there are many things we are doing wrong compared to the past generation. It was rare to find someone sitting around all day watching a football game or playing a sport. In fact, it is still rare to find someone going to college on a sports scholarship in a foreign country because a sport is not an academic subject. Instead, people worked very hard to make a life: cooking, building, cleaning, farming, and creating. In contrast, sports existed only as a minor hobby in the past, and if it were a job it was very poorly paid.

The leisure time afforded by modern conveniences (washing machines, irons, vacuum cleaners, grills, etc.) could be used to be productive and creative. The genius within each person could be harnessed into STEM fields in preparation for the next wave of Einsteins and Newtons. But unfortunately the time is wasted on "amusing ourselves to death," as Neil Postman described in his book. As stated, over 40 hours of media time and 12.6 hours of sports time per week are spent by the average high school student, with less than 1 hour devoted to science study time. This equals more than a full-time job of "entertaining ourselves." Is 52.6 hours per week what it takes to keep ourselves amused? The obesity epidemic has clear links to this sedentary lifestyle. One might argue that sports activities are the only thing keeping the nation physically fit. How can this be so, when the nation's sports playing is at an all-time high along with obesity rates?

Obviously sports are not helping! A century ago, energy needs were much higher and the U.S. population had an obesity rate of less than 3 percent. Presently, at 65 percent, the nation is overweight or obese as a result of a sedentary lifestyle. Sports do not burn off calories in the way that work does. Sports such as baseball and football involve running only at certain intervals. If you have ever watched a football game, you know it takes endless minutes before any action happens. Running is limited to intense moments, but certainly this is not the sustained aerobic exercise for which our bodies are adapted. Over 90,000 years of human evolution drives our physiology to give off energy and create and produce. Work does that for us; a sport does not. Working on a stone wall or cooking a big meal requires continual movement: lifting, bending, and cutting, moving items and thinking. Sports activity involves a lot of waiting and watching—idle time. While some downtime in leisure pursuits is good for each of us, it is the question of extremes in time that is of most concern. Emphasis should be placed on hobbies that are productive—for our health as well as societal betterment.

A Century of Productivity

The generations who built this country over a century ago, with infrastructure, weaponry, transit systems, medical advances, and building and engineering structures, are long gone. Perhaps I miss their values; perhaps you do as well. It was a society built upon hard work and a valuing of intelligence. As discussed in the last section, one had to struggle constantly to survive, preparing and obtaining food, keeping warm, and finding employment. A college education and professionalism were valued because it was rare to have an education beyond eighth grade. Recently, life became much easier on a day-to-day basis. However, in association with this, the values of hard work and respect for intellect declined.

To show this changing trend, a recent study was done to identify trends in work ethics changes and attitudes toward cheating the government. People born in more recent decades were more likely to accept the cheating behavior as OK. For example, individuals born in the 1930s were more likely than people born in the 1960s to answer that it is "never justifiable" to cheat to obtain government benefits. Younger respondents were even less likely to answer "never justifiable."[4] Quantifiable studies are limited in showing changing attitudes toward work. It is difficult to test individuals of the past decade due to their deaths and elimination from the workforce due to age. However, the vast accomplishments of the Silent generation (those born before 1930) in building up the nation show some differences between then and now.

People from a century ago who were farmers, for example, had to make their own meat by raising animals and preserving the meat. They had to know how to cut wood and have the physical strength to do it. Caloric intake was limited because of a lack of food. In fact, I recall my grandmother (born in 1902), a former farmer, who had me pick up meat for pot roast by the butcher in Queens. She would tell me, "Make sure there is a lot of fat on that meat." In her time, the little fat one obtained kept one moving and working. If ever you do manual work as they did a century ago, you'll note that the very flesh is pulled off your being. The hard work required to accomplish tasks without the use of heavy equipment and engines is intense. And there is a sense of pride in the accomplishment. There was no obesity epidemic.

Work creates a kind of solidarity which binds people together when they are creating something. It is a sense of pride that is received when a group unites to plow a field or build a house. This is probably a reason for the success of the Habitat for Humanity program. This form of collaboration is found in science labs across the world, wherein little societies bind together to do science. It is the values of former generations that are emblematic of attributes needed for STEM achievement.

In 1913, few people went to college, and the science rigor required intense study and few breezed through. Have you ever seen a textbook from that era? It is laden with knowledge and has little helpful imagery. It is complex and difficult. I do not advocate for this—in fact, it is poor pedagogy—but it shows how hard one must have had to work to succeed in STEM coursework.

There was a rugged individualism that is important in propelling science forward. Without some people willing to think outside of the box, science cannot progress. In examples of Einstein and Newton, Copernicus and Johannes Kepler, consider that each thought differently individually. This made their ideas great. Their contributions were based on skepticism and independence, which developed through their hard work.

It is interesting commentary that immigrants currently contributed to large portions of the STEM workforce. Immigrants have the drive to do STEM work, having been raised in a non-U.S. economy. Instead of deporting individuals with those skills, we should be grateful that they are willing to contribute to our STEM efforts. We should look inwardly to determine why STEM talent needs to be imported. What is it about our culture, with so many people and so many resources, that hinders homegrown STEM development?

The many facets of our society that poke fun at intellectualism and intelligence work to undermine our economy. With dominant cultural norms that ridicule STEM attributes—intelligence, science interest, personal interest in academics, and a desire to learn—it is no wonder that STEM achievement is failing in the United States. The most shocking observation is that anti-intellectualism is tolerated in our schools. In the classroom and by school administrations, participation in sports is revered and "smart" kids are often alienated. I have seen this over and over as both a student and teacher. How popular is the science club or the chess club in schools you know? How popular is being on the football team? With these differences, it is no wonder STEM has problems in the United States.

It is quite in line with the goals of the media-educational complex. I contend that it is a purposeful program to keep the populace inaccessible to the academic world and relegate them to low-paying jobs and little power in the society. A person giving up school is doomed to a low-paying, low-security job.

As a result of making it popular to view STEM and intellectualism negatively, many potential successes were turned off from academics. These students do not want to be perceived as "weird," so not studying and goofing off is supported. A school will never state that its mission is to seek such an outlandish outcome. However, it is an unintended consequence because it is tolerated. Media, sports, and socializing all often exclude intellectual students. Through tolerating these cliques, the educational structure supports an anti-intellectual stance.

Often, the excuse is that not everyone is intelligent enough to do a STEM career, so they need sports and the media as outlets for developing their skills. Sometimes Gardner's multiple intelligences are cited as a reason to develop these other areas of talents in students. Critics claim that someone who is good at sports and is poor in school needs the confidence-building experience of the sports team; otherwise she or he has nothing.

These are excuses for the alienation of the academic and elevation of the sport. While many extracurricular activities do develop personal areas (it adds exercise to one's day and has the benefit of socialization), it is the *emphasis placed* upon the activity that is the problem. A game fills the parking lot at a school, but a debate or science fair barely fills a row of cars. It is the focus on nonintellectual achievement that threatens STEM development and, in turn, our national success.

Our cultural influences inhibit STEM preparation to such an extent that we need to import labor from other countries. Outsourcing STEM talent is a mistake. A cultural and educational shift from denigrating to elevating academics must take place before STEM can be more homegrown. This will allow talents to rise so intellectual abilities remain integrated within our culture.

No one wants to be an outcast; it is a deterrent to developing academic abilities. We need an educational system that supports the socialization of high achievers into the dominant culture. Individuals who are applauded for their academic achievement are more likely to go into STEM fields.

Economics and STEM Recruitment and Retention

As stated in the previous chapter, the next great innovation, whichever nation develops it, will drive that nation to the economic forefront and geopolitical dominance. Without a move into these areas and with a national reliance on foreign nationals to help fill the void in STEM employment in the United States, our outlook is quite grim.

It is true that there were always "fame-and-fortune" careers that paid well: kings and queens, robber baron businessmen. But a century ago, the public did not delude itself into thinking that it had access to the rich and royal. People respected education, especially because so few had it and they knew it was the road to success. Intellectuals were much respected as "gentlemen" and "ladies" and society looked up to these images. Today, the anti-intellectual is elevated.

The Way We Were

This deemphasis on intellectualism and academics is a modern phenomenon. A century ago, time was spent working and few reached a higher education, so it was coveted. Education was a commodity which drew great honor. As discussed, productivity was a necessity in the past for survival. Values have changed but not for the better: alongside sports and entertainment is a loss of productivity and an unhappiness that goes with a lack of accomplishment.

When we a look at the way society was a century ago, the changes become concerning. While STEM is applauded for its accomplishments—as having improved society and human life—some changes have unintended negative outcomes. The rise of rapid transportation, communication and information transfer has been concurrent with a fall in the need for hard work. Socialization skills are in decline because people are texting and emailing without the work of communicating face to face. This requires a different set of skills, with body language, voice intonation, immediacy in responses, and clarity needed to convey and understand thoughts. Use of technology is moving society away from solidarity and into isolation.

The very aspects of our culture that built its infrastructure—dedication and solidarity gained through labor—are in decline. People are on the computer and use media devices for more than 40 hours per week, detracting from their time with one another. A whole set of interpersonal intelligences gained through human interaction is diminishing. These attributes are all necessary for a successful academic pursuit into STEM fields.

Technology is using us in more ways than we are using technology. Spending time on social networking sites limits time doing yard- or housework and studying for a class—in essence, being productive. It is true that technology—the bulldozer, the computer, the crane—has improved an individual's capacity to get more done than an individual's human labor of the past. However, a misuse of technology is drawing people away from creativity and workmanship, which always give a sense of self-satisfaction. Technology removes the difficulty of doing tasks but with it the experience of struggling through work and the gratification of accomplishing work diminshes.

Immigrants are picking up the slack by doing the labor at cheaper costs than homegrown workers. Lawns and carpentry are being done by illegal immigrants for many reasons. Their labor is cheaper and, in accordance with supply-and-demand principles, they will always get the jobs before our own businesses. In STEM areas, if immigrants are the sources of labor, then those individuals will garner the profit from that industry. We only have ourselves to blame if STEM is dominated by immigrant talent. As a society, if we do not choose to increase homegrown human capital in STEM fields, it will be to the loss of our citizenry. With skilled labor in short supply—over 600,000 jobs in STEM fields remain unfilled—it is surprising that more people are not entering these careers.

In fact, there is increased interest in allied health and STEM fields for the reasons given. But the rates of actually gaining the skills and degrees to participate in that part of the workforce are low (between 20 and 45 percent). The interest may be there, but the ability to do so is not. We have the same abilities as the generation a century ago, but those talents are being thwarted.

The "way we were" was to go to school, develop skills in an area, and get a job. The "way we are now" is to go to school, get lost in myriad social activities, *not* develop the right skills, and *not* get a job. Greater emphasis on tracking people into STEM fields which have high demand is a necessary precursor to improving STEM achievement and lowering unemployment. Retraining workers who are made obsolete by technology into STEM areas is essential. Programs to move population from one field into another is based upon people's belief that they can do it. It is based on their self-confidence that they are able to do STEM work. After years of education in which people are socialized to make fun of or alienate people who are intellectuals and science nerds, it is difficult to make those changes.

However, a shocking transformation happens. The very unemployed who participated in the alienation process of the intellectual are now faced with intellectuals as their potential bosses. There are many sitcoms about this twist of fate: *Frasier, That '70s Show*, etc. For example, a former nerd in high school is now the CEO of a major firm, employing the former high school football quarterback. The reality is that many people are capable of contributing to STEM aspects of the economy. Few are able to make it through the cultural and educational obstacles in place along the way.

Collaboration

The success of the U.S. STEM sector of the economy will require cooperation by several elements of the culture. The media-educational complex, governmental agencies, and economic structure will need to sell STEM careers as both attainable and cool. The workings of the culture will need to create an image of science as something to be desired, with people working together seen as patriotic and productive, bolstering the United States to dominance.

The opposite has occurred: the contemporary shift from collaboration to physical competitiveness clearly contributes to a national decline in skills needed for STEM achievement. According to the many years of TIMSS (Trends in International Mathematics and Science Study) reports, the United States has steadily fallen behind other countries in science and mathematics scores, from first place in the 1950s to last among the Western nations today.[5] One cause may be that other cultures emphasize closely working together in a non-sports capacity: in industry, Japan has collaborative shareholding among its workers, and European schools have vocational programs that include early adolescent apprenticeships in trades, requiring productivity and collaboration. Emulating some of these strategies should help foster the kinds of collaboration needed for success in U.S. STEM areas.

Collaboration (and not competitiveness) should be emphasized for creating a culture successful in STEM areas. Science is a community, and intellectual collaboration is a key to its success. It takes a group to work together; sports teams may foster such comradery. Religion has elements that mirror those of scientific communities. Both have rules based on honesty and collaboration. Both seek truth and epistemological answers to ultimate, large questions. Both seek to educate and change society. However, science is based upon reason and the process of objective discovery, and religion is based upon belief. Science can draw from the elements of religion to improve its development. Religion can use scientific process to gain understanding of the natural world.

Other Diversions Deterring Motivation

There are social and cultural diversions in addition to sports; entertainment/media and technology (e.g., gaming) diversions also hinder academic achievement in science. Omitting the importance of time needed to prepare for a STEM career, it is motivation that matters more. It is not merely the hours spent on long-shot careers but the integration of intellectualism within the family and the community. If the dominant culture deems nonscience topics, such as sports conversation and entertainment knowledge, as valuable, then a shift in focus from academics will reduce motivation.

STEM conversation should be intrinsically valued by the culture and within the family. In a peer cohort, STEM discussion should be seen positively: talking about health issues and how the body works, and talking about nature and how animals behave, is vital to making science a part of the mainstream. These topics, as discussed in previous chapters, are intrinsically interesting to people of all ages. It is their unglamorous image that suffers from a media-educational complex that artificially elevates sports, entertainment, and media-technology as more fun and interesting. It is not true, but it is being sold to the public as such. This is one myth that deserves debunking—that STEM is not intrinsically cool.

The lure of a media-enticing career in entertainment or sports is based on passing fads. STEM concepts are lasting and represent what is termed in German *die unsterbliche musik* (the undying music) of life. Our human functions, the way the Moon orbits the Earth, the patterns of weather, and the rise of the sun are STEM concepts that people will ask themselves about for all posterity. But who will ask about Angelina Jolie and Eminem?

STEM people are remembered, and their accomplishments change the world, as seen in the last chapter. People in popular media careers are hardly remembered. Very few movie stars are remembered after their very fleeting fame is gone. When you consider that very few people still remember Barbara Stanwyck or Joseph Cotton, former movie stars of the 1930s (although I like Joseph Cotton), it is clear that these low-demand areas only offer short periods of fame for even their most successful entrants. Everyone knows of Albert Einstein and Thomas Edison. If people do not remember these great scientists, they certainly know their creations—radioactivity, light bulbs, electricity—and their future applications. It is true that science touches eternity but that the lights of Hollywood fade fast.

Efforts to retain and recruit qualified people into science are threatened by pop culture. Many people follow trends and forego a STEM education. These passing fads are one of the great lures created by the media-educational complex and its de-emphasis on science. It is natural to want an instant fix or a quick career. Long years of struggle in academics sound frightening. However, if these years can be enjoyable learning, then students would be lured by such a reputation. Science jobs could have an environment which appeals to students.

II. CAREER (SCIENCE EMPLOYMENT STRUCTURE)

Science Employment Itself as a Barrier

Science jobs often have environments in which a laboratory setting is isolating and unpleasant. Doing work with chemicals or microbes in an enclosed room without windows is a real image of science employment. It may attract asocial personalities or people who can withstand long hours with a focus on procedures. This is a real description of many forms of science employment.

Allied health careers also require environments which some may consider unpleasant: emergency rooms filled with high stress, blood and gore in accidents, depression from end-of-life scenarios or extended care of the elderly. Some consider each of these cases invigorating and their career choices are in line with the environment. Allied health work climates (just as other jobs) attract a certain personality who chooses the environment for a reason.

There are, however, many varied STEM work climates: teaching and health education; outdoor work with infrastructure, land, water, or animal studies; medical treatment that involves all types of less-stress and less-gore climates; laboratories with social and physical facets in progressive laboratory industry (e.g., child care and gyms, along with worker education). These examples show that publicity of these alternate images would have an impact on STEM public relations. Many times I hear from students that they would enter into a science, but they are too social or don't like the gore of health field careers. If the media-educational complex would devote more time to the alternative career choices in science, people would become more aware of their options.

Critical Thinking as a Barrier to STEM Careers

The largest challenge to science progress in the U.S. is a lack of critical reasoning ability. Critical thinking is defined as an ability to work through problems and ideas and develop logical conclusions. The phrase is often used interchangeably with terms such as critical reasoning, scientific reasoning, scientific thinking, and rational thought.

While critical thinking is lacking in the national ethos, as discussed in previous chapters, it can be overcome by strong cultural and educational forces to bring people up to the standards required in a STEM field. To be clear, a lack of critical thinking is the main roadblock to science literacy and to success in a STEM area for most people. Development of this area of knowledge is a key goal of this book but also for science education along the pipeline.

Without the ability to reason at higher levels in the typology of argumentation, understanding in science will be limited. It is vital to point out that science literacy is inherently linked to mathematics literacy. Speaking from experience as a pre-allied health advisor, the insurmountable barrier to entrance to an allied health field is lack of mathematics competency. Many times students enter programs that require a course in college algebra. At that point, I realize that their career path looks grim because the prerequisite knowledge of algebra and abstract thinking that accompanies mathematics learning is lacking in these students. It is hard to believe that one could go through an entire secondary program and not have algebra. It is even more disappointing to fathom that a high school teacher or advisor did not inform the prospective student that a degree in allied health would take well more than five years to accomplish without basic mathematics skills.

Usually one-half of entrants to any major's general biology course in college are premedicine (or some form of allied health). Over half of these students fail out of the course or change majors after the year's end, nationally. There are individual issues, for sure, as described in the algebra example given. There are also serious structural contributions upheld by science education that create an unattractive environment for students considering science.

III. SCIENCE LEARNING ENVIRONMENT

Science Education

People will accept the cognitive work behind a STEM discipline if the science education environment would make it an enjoyable experience. That is not to say that there should be fun and games, but a cold and uncomfortable environment is also not good for engagement. Science has a great deal to offer students.

There are aspects of the science education curriculum that could be very appealing. It is a free-thinking domain that fosters individuality and reason, both innate interests to human beings. STEM presents a curriculum that is clear and ordered, with step-wise development from one course level to the next. Topics include our own bodies, our own chemicals, and experiments to test and see results. All of this appeals to our senses. It is also a very lucrative area, in high demand economically. Most of the top ten career choices based on salary are in STEM areas. This publicity is good for STEM recruitment. Indeed, interest is at an all-time high for many STEM areas. Then why is there still poor retention and recruitment into the majors and careers?

The PR (public relations) image for STEM education is quite bad. STEM has a poor standing in the larger society because there are facets of science education that turn off students. The focus of the next section is to 1) discuss the elements of science teaching and learning that contribute to dissatisfaction among its student members and 2) to suggest changes and modifications to improve the attraction of a STEM education.

There is a good reason for the bad publicity science education receives: STEM retention rates are abysmal. Less than half of students entering a STEM major actually finish within six years. People know this when entering college and expect the worst. However, better preparation and a smoother transition to college would ameliorate this problem. The outcome for this chapter should be an analysis of STEM education and appropriate recommendations for improvement.

Improving Science Education

The state of science education is in jeopardy because education and intellectualism in the United States has declined. It is in part, the fault of the larger culture, as discussed in the previous section. It is in part, due to poor science teaching and curricular structure. Statistics show students lagging behind other Western nations in all academic areas, but worst in mathematics and science. Suggestions for change in the dominant culture have been made and are essential for STEM improvements.

The issue is complex, but to start, an increase in the standards for teachers would help gain teacher respect. Because many science teachers are also not properly certified or qualified to teach science, this would ensure good knowledge in the faculty. Current teachers would need further education if they did not have a major in their field of study. The requirement of a master's degree is debatable at the secondary level, but certainly each state should place resources into changing requirements and supporting those teachers to follow suit.

Standards for enrollment into STEM majors should also rise, along with financial compensation. Again, resources need to follow these increased admission standards. Scholarships for STEM students, improved assistance for tutoring, and student-faculty research have been shown in the education research to improve retention. Not only would changes in financial packages for STEM students make the field more competitive, but it would attract better students from other fields. Currently, STEM areas are the only majors that lose more students than they gain through the educational pipeline. Recruitment of quality students takes resources that government and institutions, if they are serious about the issues, will find a way to fund.

Ingenious methods are being employed in progressive institutions: select grouping of STEM students in dorms to build collaborative teams; advanced faculty-student research to improve bonding between the two and improve retention; and enhanced research at earlier stages of a student's experience to help him or her practice doing science like a scientist. All of these are examples of methods of structuring science education to improve recruitment and, more importantly, retention in STEM majors.

By accepting only the best academically prepared students into science teaching, science will gain an improved reputation in the dominant culture. STEM teaching and learning would become more ingrained into the society as something to be valued. That teaching is to be respected and that science is an entrance into a valued field of study would recruit and retain more students. Better academic standards for teachers should be supplemented with higher wages to both boost prestige of the academics and attract an elite force of teachers.

Unfortunately, the argument on the other side points to a lack of resources to pay teachers, especially at a time of tight budgets and across-the-board cuts. One needs only to look at where the money is being spent and where it is not being cut: the military, wars, waste, building sports stadiums (a new Yankee stadium), and diverting resources to foreign operations overseas, including grant money to influence other nations. These are only a few examples of priorities for which the governmental system has plenty of public cash on hand.

These principles apply to STEM because it is a major force in a global economy. We are quite interdependent because of open trade with one another. Of course, this leaves nations vulnerable to one another's economic problems. STEM areas continually supply new innovation to this network. Empowering STEM areas can be the answer to our economic woes. The weakness in our current state of science education is an obvious contributor to the economic downturns we are experiencing.

Science Teaching Reconsidered

The goal of this section is to explore the aspects of science teaching and learning that strengthen and weaken retention in STEM classrooms and programs. It is well-documented that too few college students are recruited and retained in science programs to meet the nation's future needs.[6] The National Research Council reports that first-year student retention in undergraduate STEM majors has remained poor over the past decades.[7]

A study conducted by UCLA's Higher Education Research Institute found that STEM students take longer to complete their degrees than non-STEM students. The study tracked thousands of students who entered college for the first time in 2004.[8] More specifically, first-year student retention in STEM majors is at an all-time low.

The Center for Data Exchange and Analysis followed students who entered STEM bachelor's degree programs in the 1993–1994 academic years and found that only 38 percent of those students earned a STEM bachelor's degree within six years. There were also disproportionate graduation rates between white and Asian American and URM groups (underrepresented minority groups: Latino, African American, and Native American). Disaggregating this data by race, 23 percent of URM STEM students earned a STEM bachelor's within six years and 41 percent of white and Asian American STEM students earned a STEM bachelor's degree within six years. When comparing four years of college completion rates, the percentages dropped considerably—by almost half for most groups. Completion rates for 2004 freshmen STEM degree entrants who in 2008 (four years) follow: white and Asian American groups had completion rates of 24.5 percent and 32.4 percent, respectively, and Latino, black, and Native American had completion rates of 15.9 percent, 13.2 percent and 14.0 percent, respectively. On a more positive note, the proportion of students reporting interest in STEM-related majors remained consistent between 1971 and 2009.[9]

The overall numbers are not pleasing, however. With low rates of retention, student interest cannot materialize into STEM careers. Overall, the low rates of students reaching their goals within a reasonable time frame indicate systemic problems in science education. What factors are contributing to students' leaving the STEM major?

In addition, URM group retention differences result in the loss of disproportionate amounts of talent from these groups. Are there aspects of the STEM learning culture that contribute to these losses? It would be prudent to explore these elements and to develop institutional and curricular changes with these groups in mind. Retention is a national issue and remains a major concern for any science program, and such numbers should be an impetus for change.

Why should a reader be interested in science education retention? The prosperity and safety of the nation depend on human talent entering into the sciences. Economic, military, philosophical, through practical progress has always been made by human scientific thinking. Without human talent attracted into science, the nation and world weaken substantially.

The Great Divide: High School versus College Science

Students drop out of STEM courses and programs throughout the educational pipeline. However, a particularly weak transition occurs between high school and college science for students. Incoming first-year students expressing an interest in STEM majors have an overall student loss rate of 40 percent at this juncture. Losses range in STEM majors, with 50 percent in the biological sciences to 20 percent in the physical sciences and mathematics in the first year.[10]

The highest rates of dropping out occurred for high school graduates who withdrew their decisions to enter a STEM major at or before enrollment in college. However, during college, the highest risk of STEM switching (35 percent) occurred at the end of the first STEM year. Some students dropped out completely but most entered a non-STEM major. Retention rates improve dramatically as students move through the STEM major. During the sophomore and junior year of college, roughly only 2 percent decide to leave each year after the first.[11]

There is always a loss of students from STEM areas to other areas of study. STEM areas have the highest defection rates and the lowest rates of recruitment. This may be because of the vertical structure of the STEM curriculum. One cannot simply transfer into STEM without prerequisites being completed. As discussed earlier, pre-allied health majors need a substantial amount of preparation in mathematics to be successful in science. Nonetheless, the difficulty in recruiting new talent into STEM areas throughout the college years contributes to science's poor retention rates.

Institutions could provide additional support structures and more flexible scheduling to accommodate STEM entry at various points of the academic career. An outward flux of students results in the loss of talent; but again, it will take resources (tutoring, extra courses, trained counseling, and advising) to bring those talented students from outside into the STEM major. Medical schools recognize non-STEM major value and give preferential treatment to non-STEM majors to add alternate talent to their group of aspiring medical doctors. The idea is that a music or art major will have a more well-rounded training for working with patients. Of course, the prerequisites of mathematics and science courses must be completed before entry into a medical program.

Career Entry Losses

While retention rates have remained low over the past decades, career entry into STEM-related fields has also suffered. Many areas have obviously been affected by poor retention rates, resulting in a shortage of qualified health care professionals, engineers, science educators, and trades. Skilled workers (teachers, in particular) in STEM areas will leave their respective fields, with a STEM degree in hand, to pursue more lucrative industry jobs. Why work for $42,000 in academia with a PhD in geochemistry if a petroleum company will offer you $120,000 to start? Why deal with the increasing funding competition in academia and teaching when there is flexibility and free time in industry as well as triple the salary? The answer lies in individual interest and commitments. However, that is, again, not good enough. STEM-related areas in industry pay more and educational institutions need to be on par. Otherwise, the more talented will go to industry jobs, and the weaker STEM graduates will be the nation's future teachers and professors (albeit vast generalizations).

A shortage in the supply of science teachers in many different geographic areas was specifically noted by the National Research Council. However, the reported declines of 60 percent in the number of students preparing to teach science are also alarming. Although a variety of factors may contribute to STEM teacher scarcity, clearly the poor retention rates for STEM college majors is playing a role.[12]

Low STEM retention is linked to declining scientific literacy in the public as a whole. Science literacy is the ability to know and be able to do science in the ways described in the definition of it presented earlier in this book. Low retention means that fewer people are being reached by traditional STEM education. Public knowledge about new innovations in science is largely accomplished through the media, which is concerning. Poor retention has thus produced a nation that has "simultaneously and paradoxically both the best scientists and the most scientifically illiterate young people. . . ."[13]

Declining scientific literacy, combined with the decrease in STEM graduation rates, has also resulted in reduced numbers of people in research development, a driving force in the progress of science. Public concern has thus been expressed regarding the international competitiveness of the United States in the science and technology-dependent sectors of the economy. However, it is an international crisis because U.S. innovation has traditionally driven STEM and the world economy.[14]

Thus, the major purpose of the next section is to explore the empirical evidence about science teaching and learning that may link to poor retention in STEM areas. The hope is to make recommendations for the current educational system. Also, by being informed about the current state of science education, readers can be better prepared for a career and education in science. Finding ways to improve STEM education for all students is the goal of this chapter. Socioeconomic effects of STEM deficiencies on U.S. competitiveness will be explored in the next chapter.

The State of Science Teaching

University faculty have traditionally explained undergraduate attrition from STEM majors as appropriate, claiming that those who are unprepared and lazy are weeded out.[15] This may well be true, but what about the qualified students who leave? This section gives the latest science education research through a systematic exploration of each element of STEM education. A variety of studies have been conducted in order to tease out the variables that contribute to a student's decision to switch from a STEM major: Loftin in 1993; Razali and Yager in 1994; Strenta, Elliott, Adair, Matier, and Scott in 1994; Seymour and Hewitt, 1997; McShannon in 2001; and Daempfle in 2004.[16]

Climate Change in STEM Classrooms

The STEM classroom environment is the most important influence on learning and attitude among students. Differences between STEM and non-STEM classroom structure were found to contribute to student disaffection with science. Student loss rates were directly related to an unwelcoming classroom climate.

A major study by Elaine Seymour and Nancy Hewitt from 1990 to 1993 examined various aspects of STEM education that influenced students to leave their majors. The overall aim of the study was to analyze influencing decisions of high-achieving college science majors to switch from a STEM major to a nonscience-based discipline. The key is that they looked at only high achievers. Losing talent is a concern among all STEM educators.

Seymour and Hewitt conducted an ethnographic study with interviews from 335 students in STEM majors drawn from seven major university campuses containing a high proportion of these majors. Most of the data were gathered through personal interviews and in-depth analysis. An additional 125 students participated in focus group interviews on six other campuses. All of the college STEM majors were considered well-prepared, having math SAT scores above 649. Half of the students included *nonpersisters* (those switching from STEM majors), and the other half included the *persisters* (those remaining in the STEM majors).[17]

In a surprise result, Seymour and Hewitt reported the same set of concerns in both groups. Complaints by both persisters and nonpersisters showed serious discontent about the quality of teaching in STEM courses. The data showed that nine out of ten nonpersisters (90 percent) and three out of four persisters (75 percent) described the quality of *science instruction as poor overall*. The climate of the classroom is described as "chilly," with distant and uninterested STEM faculty lecturing and uninvolved. Students cited issues about professor effectiveness, inappropriate testing styles, and inadequate course availability. Students strongly believed that science faculty did not like them and, moreover, did not like to teach. The STEM faculty gave the impression that they did not value teaching as a professional activity, but instead valued their research over teaching.[18]

In contrast, when asked to compare STEM courses with non-STEM courses, students expressed the following set of antonyms: coldness versus warmth; elitism versus democracy; aloofness versus openness; and rejection versus support (STEM descriptors were the first in each contracting set of words). In interviews and focus groups, the most common words used by first-year students to describe their personal encounters with STEM faculty were *unapproachable, cold, unavailable, aloof, indifferent,* and *intimidating*. Students also described the STEM classroom as based on sarcasm and ridicule, put in place and tolerated by faculty.[19]

Seymour and Hewitt exposed a nasty side of the college STEM classroom. Obviously, there are excellent teachers and great classroom in STEM majors throughout the nation. However, empirical evidence indicates that norms are not so friendly to students. These practices, rarely found in non-STEM courses, are described by students as discouraging. The STEM classroom disregards student personality and contributions. Students describe voluntary student participation as frowned upon by STEM instructors and cite an atmosphere of intimidation. Nonpersisters cited these issues as a main cause of their decision to leave the STEM major. This phenomenon in STEM classrooms is known in the research literature as *the chilly climate hypothesis*.

What, in particular, is it about STEM norms for teaching that allow this environment to continue? In studies, criticism from students focused on a lack of discussion in the college classroom. STEM students and former STEM students describe a one-way transfer of information through traditional lectures. At the same time, students valued their high school experiences, in which they describe a completely different approach to teaching. High school science classrooms had open discussion and a good deal of dialogue compared with their college lecture courses in STEM areas. Students also cite poor quality of preparation for classes by instructors, a focus on rote memory as a class goal, and even faculty reading directly from textbooks. These were described by students as factors contributing to poor STEM instruction and their eventual decision to leave the major.[20]

A common explanation given by college science faculty for poor retention blames the students. Often, poor high school preparation is pointed to as a reason for student dissatisfaction. To some extent this is obvious, but simple dismissal of large numbers of students is inappropriate. How do students get prepared for college STEM properly? These next sections will discuss methods institutions and faculty can employ.

While high school preparation is not directly to blame for these issues (secondary teachers cover material they believe will be necessary and useful for the science student), secondary teachers have some responsibility for not preparing students for STEM success in college, as the expected college skill set and believed preparation often do not coincide. However, without knowing what to expect, secondary schools are left in the dark. This chapter is a first step in clarifying what is to come for their students in college.

Students in a variety of studies appear to value their high school experiences in science and math over the college STEM courses, citing better instruction and a more interesting learning environment. This indicates a difference between high school and college STEM teachers in what they expect of a classroom and of their students. Because students favor the high school situation, the attributes of college classroom should be reconsidered.

Could nonscience subjects simply be easier? Is STEM instruction getting a bad name only because it is tough? Researchers Strenta et al. (1994), Seymour and Hewitt (1994) and Gainen (1995); published research in which they asked students to elaborate on their reasons for leaving a STEM major. Participants expressed their general unhappiness in STEM areas for similar reasons as found in the Seymour and Hewitt study. Results showed that students were generally interested in the sciences but were "turned off" by the structure and climate of the college STEM classroom.[21]

There have been complaints about language problems with foreign assistants, large class sizes, and poor high school instruction in several studies. However, an intimidating classroom climate, poor quality of undergraduate science teaching (particularly dull lecturing), and a general lack of nurturing for the student were the most frequently cited concerns among STEM students. Even students remaining in the major recognized that non-STEM courses were more interesting and more pleasant than STEM courses.[22]

There are a number of changes that should be made to the STEM classroom climate in response to these research findings. First, a move to more collaborative and discussion-oriented classroom structure should be advocated by institutions. Second, incorporating cooperative learning strategies instead of competition would help develop peer support networks. For example, in my laboratories, I use methods of group assessment and group participation in answering questions and performing lab procedures. I use a "gambling for grades strategy" in which groups work together and bet points in their exit assessments. Students get to know each other and form study groups through natural interactions in the setup. It works quite well with some restrictions and a little bit of bravado from the instructor. Strategies are unique for each instructor—what works for some may not work with the personality and preparation of another. Regardless of the particular strategy, the goal should be to improve student interactions and dispel the chilly climate. A third solution to the current climate problem should be to ask questions and more questions, limiting lectures and moving to transmitting information through a Socratic method. For example, content learning in this book was accomplished through debunking myths and not from a series of standards addressed in lecture notes. Content was infused in discussion throughout this book to address all of the standards advocated by AAAS (the American Association for the Advancement of Science) and NSES (the National Science Education Standards). It should not have "felt" like lecturing to the reader. Similarly, content should be learned through infusion into creative discussion and with societal interests.

It is important for readers who may be considering a STEM major not to be dismayed by these studies. Through becoming more aware of how to make STEM education better, a student's entrance into a STEM field will be smoother. This should help in understanding how to avoid pitfalls in the current STEM system of education. There are many progressive schools and instructors attempting new instructional strategies in STEM areas. New research will emerge and STEM education will change. Until that time, students entering into a STEM classroom should be prepared to meet the needed requirements or make adjustments to succeed. Perhaps by getting more tutoring or by developing STEM peer relationships early on to work in study groups, students will have better chances of success. This chapter and its background research give a guide for improving STEM education.

Personal Contact with Faculty

The biggest changes to student attitudes about STEM education occurred when they were allowed to engage in research with a faculty member. College science is about gaining an understanding of the scientific method of thinking. Students crave this aspect of science. Research can be learned through each chapter of a book (like this one) but it is enhanced most when it is practiced. When students were allowed to participate in faculty research, they became more involved in class discussions. However, few students were able to experience such a relationship in the studies reviewed. The few students who had experienced it expressed that they valued the open relationship with faculty in a research situation compared with faculty's apparent indifference to them in a lecture hall.[23]

Mentoring experiences, as just described, are strong predictors of persistence in STEM majors. Regular personal contact with a particular instructor who took interest in the students, departmental gatherings, and small group learning and discussion all contribute to creating a warmer environment and retaining students. It is clear that many students want to remain a STEM major but are turned off. Through increasing faculty-student interactions outside the classroom, inside classroom climate will improve. Of course, high student-faculty ratios limit these kinds of attempts. Again, resources need to be committed to get class sizes smaller and compensate faculty for doing student research.[24]

Another frequent complaint among STEM students in several studies was inadequate personal contact with faculty during advisement. Too many students per advisor lead to poor advisement and poor retention in STEM. In fact, many students traced their retention problems to poor or improper faculty advisement. A large majority of STEM students (one-third of both switchers and non-switchers) felt that they were not made aware of the length of study (e.g., more than four years) required of STEM majors.[25]

Women and Underrepresented Groups in Science

STEM remains a field that disproportionately retains students based on gender and race. As discussed earlier, URM STEM majors have almost twice the dropout rates as compared with white and Asian American groups. Engineering and physics, in particular, remain the most exclusively male-dominated STEM disciplines. Data show that minority women disproportionately fail to be recruited and retained in engineering and physics, with first-year enrollments and retention declining. The study blames a feeling of student isolation within the engineering academic programs. More recent studies of engineering students found that minority women were more successful in nontraditional learning patterns (e.g., cooperative learning styles) than their cohorts.[26]

The strategies recommended in the previous section would help ameliorate this cause of URM group dissatisfaction. Retention among underrepresented groups in STEM areas would benefit from changes in instructional styles that include multiple learning style approaches and a departure from the competitive "chilly climate."[27]

The largest study to date, funded by the Sloan foundation and for which I am an active reviewer, is set to explore the nature of retention in engineering. The study analyzes a large segment of engineering programs and its results will publish in 2013. An important goal for the future of science is to retain women and URM groups in science to add their talent and expertise to the field. Further research into the contributing factors behind their poor retention rates should be encouraged.

The Great Divide: STEM Changes

High school and college educators do not agree, on many points, about what is required of incoming college students for success in introductory STEM courses. It is surprising because both have had the experience of their own STEM high school and college days in the classroom.

This may be a contributing factor. People without training in a STEM field often teach the ways they were taught. Thus, the system of teaching perpetuates itself. However, times and strategies change so that both sides of the divide between high school and college instruction remain in the dark as to what the other is doing. College teachers design their introductory courses with certain ideas in mind for what students should know. High school teachers teach to prepare students for success in college STEM courses.

Figure 11.1. Great Wall of China. Sources: iStockPhoto/ThinkStock.

According to a number of studies, my own included, there is a mismatch in preparation at these levels. My research showed that college STEM instructors had very different expectations for incoming students than those expected by high school teachers.[28] Differing expectations and thus preparation between high school and college may lead to a great divide in the transition, as shown in figure 11.1.

In particular, my study examined how well matched high school and college biology teacher assumptions were about what is required for success in college biology courses. The study used a focus group approach so that face-to-face contact would elicit ethnographic data. The results of this study indicated that the two groups were very much unaware of one another's ideas. College teachers emphasized the importance of mathematics, writing skills, and integrating biology with other subjects, for example. These were non-biology expectations for knowledge that made secondary faculty very upset. They had been preparing students to learn biology. The focus group experience had completely thrown them off their belief system. One of the secondary teachers walked out, angry that these differences were insulting to his work over the years.

It is tragic that high school teachers who want to prepare their students do not have sufficient contact with current college curriculum and instructors. Communication would help a great deal, which is why I structured the project in such a manner. In fact, the high school faculty instead valued other content areas such as vocabulary knowledge and nomenclature skills (e.g., Latin usage), and dispositions such as self-discipline. These were expectations from their college days and were no longer applicable.[29]

The results imply that high school teachers, who believe that the content they teach is important in their students' preparation, are, in fact, concentrating on areas that college professors do not value highly. This mismatch of standards and guidance is the fault of both sides and college STEM courses are suffering along with student retention.[30]

Epistemological Assumptions

How people view knowledge is important in determining if and how they will acquire it in science. Evidence shows that student *epistemological assumptions*, or how they view science knowing, may affect their academic performance in STEM.

When someone views knowledge as certain and believes in the black-and-white of thinking, they are not engaging in scientific thought. This has been shown to be a barrier to success in STEM. The more students believe in the certainty of knowledge, a lower level in the typology of argumentation, the more likely they will not "get" science. Science needs to be practiced at higher levels of knowing.

These lower-level assumptions about thinking lead students to oversimplify knowledge. Einstein stated that you should explain something as simply as possible but not more than that. When students interpret science information as absolute, they lose this essence of scientific process—that of continual change in accepted science knowledge.[31]

Epistemology affects retention rates. Perhaps students with naive epistemological understandings of science may view the STEM classroom as a chilly climate because they are not "getting it." Students with lower-level, right-or-wrong beliefs about knowledge probably become dissatisfied when confronted with STEM classroom demands for higher-level thinking. It is unclear, but this may link to poor student critical thinking in science, a main barrier to success.[32]

In fact, a majority of STEM dropouts have a common way of viewing STEM knowledge. As a group, nonpersisters showed an uncritical acceptance of information as factual and right-or-wrong answers. Students did not critically consider knowledge and the plausibility of it being wrong. Most nonpersisters considered science learning the acquisition of "a thing" and a solid set of facts. They sought a product that could be obtained and used (a college degree), so the essence of science was lost. A utilitarian purpose was the reason for obtaining a science education.[33]

While I commend a focus on career, STEM majors should not enter the field simply because it is in high demand and will lead to a higher salary. This is their recipe for failure. Return to the quote by Confucius at the beginning of this chapter. A career in STEM should be driven by curiosity and excitement about science and not by money.

We began the book with a philosophical approach to science; one that attempts to bring the reader to higher levels of science thinking within the typology of argumentation. It is a goal of this book to guide the reader to view science as changing and in flux. Further, it should be the goal of science education to continually debate issues and consider alternative ways of thinking.

The Need for Change

In summary, changes in classroom structure should be sought to improve student achievement and retention in STEM fields. There are many elements about the culture, careers, and educational systems supporting STEM fields which interact to create student motivation and success.

This chapter clarifies and interprets the interaction of those characteristics that perpetuate low retention rates throughout the educational pipeline of STEM. The interaction is quite complex and many opinions lie on either side of the arguments presented in this chapter. There is no need for dissatisfaction with the economic, political, and social perspectives taken by the author.

As a scientist, this point of view should be considered with the singular goal of improving STEM education and progress. The need for changes in so many areas should be addressed in order to enhance science's appeal. The need for students entering college to understand the state of STEM teaching is vital for students and for the success of science programs.

12

Driving the Economy through Science

STEM DRIVES THE ECONOMY

Industries and jobs linked to STEM (science, technology, engineering, and mathematics) areas comprise a large portion of the nation's economic growth. It is the driving force behind our national competitiveness and is gaining in productivity in the global economy. According to the U.S. Department of Labor, only 5 percent of U.S. workers are employed in STEM fields. These same workers are responsible for more than 50 percent of our economic growth. STEM employees invent products and perform highly skilled services that change our lives and create jobs.[1]

These jobs are better paying than non-STEM jobs. It should not be a central reason for going into the field, but it is an attraction. In fact, college graduates working with associate's degrees in a STEM area outearn 63 percent of people who have bachelor's degrees in nonscience fields, according to the Georgetown University Center on Education and the Workforce. Those with a bachelor's degree in a STEM major earn over $500,000 more in their lifetimes than non-STEM majors. In the past thirty years, salaries in STEM-related jobs increased at a more rapid rate than in other areas. STEM wages rose 31 percent over the past thirty years, while those in non-STEM occupations only increased by 21 percent.[2]

Yet only a small percentage of people pursue careers in scientific areas. In fact, the rate of interest of first-year college students in science and its related mathematics and engineering majors has increased considerably over the past four decades, while the rate of actual completion of a STEM major remains poor. There are many reasons for this national trend, as discussed in the previous chapter. The purpose of this chapter is to address this problem by 1) delineating how the economic variables link to STEM issues and 2) recommending changes to improve science's advancement.

Recruiting and retaining qualified people in the sciences are major goals for any advanced society. Shortages in STEM-related career professionals are the leading cause of systemic problems in the nation: a lack of qualified medical doctors, particularly in primary care areas; deficiencies in or underprepared science teachers; a major nursing shortage, predicted to worsen in the coming years; a lack of science researchers in medically oriented fields such as the pharmaceutical industry; and few physical scientists for military technology development. High-demand careers garner higher wages, and yet fewer and fewer people are prepared to successfully enter into those areas, as discussed in the previous chapter.

As a remedy, the United States is outsourcing to meet these needs by importing STEM talent from other countries. Less than half of all STEM PhD candidates in U.S. universities are American-born students. This is shocking in a nation with eighty-eight million adults not working. With high unemployment rates and especially concerning underemployment rates, it is critical for our economic outlook to improve STEM entry.

While science is universal, an inability to have homegrown talent sustain the nation's STEM needs is clearly a threat to our national security. When science literacy is not sufficient to produce enough individuals to suit the nation's high-skill needs, concern is warranted. In addition, with STEM retention in decline, the United States is not doing its part in contributing to international science and the global economy. It is true that we are the largest economy in the world and produce the most STEM products and innovation, but, given our population and resources, deficiencies in science literacy may have negative long-term effects.

THE FALL OF U.S. DOMINANCE IN STEM AREAS

For the past half-century, the United States has operated under the assumption that it will remain a superpower forever, given its vast abundance of resources. The nation was able to win wars and outperform industrially due to technological advances and innovation by the generations of the past. But, as shown in the previous chapter, cultural and educational changes now threaten this dominance.

In history, improvements in technology and science helped any civilization to win nearly every war and every battle. For instance, the modern use of tamed horses in battle led to Asiatic victories over European tribes in ancient times. Nuclear fission ended World War II rather quickly and continues to defend our shores from foreign invasion. Rocketry, submarines, and advanced weaponry probably act as a deterrent enough to stave off many military and political attacks by foreign powers.

An obvious question is "What is the next technology to supplant our existing advantages?" One can only speculate about possibilities from Star Wars missile defenses to better technological controls of bombs. In a strange way, it will take innovations that have been developed perhaps by accident or within the mind of a genius . . . something less feasible but strategically coveted, like invisibility. Just imagine if one were able to become invisible! Then simple infiltration of our defenses could lead to the destruction of our nation. Flight seemed farfetched to my grandmother's generation at the turn of the nineteenth century. While seemingly impossible, the point is that technology and sciences are vital to our security and economy.

International testing shows several trends in our adolescents' science and mathematics scores. It is a set of comparatives that are quite disconcerting. The Organisation for Economic Co-operation and Development (OECD) reports that U.S. fifteen-year-olds rank seventeenth in the world in science and twenty-fifth in mathematics out of twenty-five countries. More specifically, TIMSS (Trends in International Mathematics and Science Study) shows U.S. physics and mathematics students scoring last among sixteen nations on advanced portions of their test. NAEP (National Assessment of Educational Progress) test scores show no improvement in science over the past forty years. This is at a time when STEM is more and more in demand. Our college graduation rates rank twelfth and our infrastructure ranks twenty-third in the world. This is at a time when STEM college degrees are more and more important for developing new infrastructure and innovations.

As pointed out in the first chapter of this book, only a few decades ago, in the 1950s, STEM areas were more respected by our dominant culture, and the United States ranked in top positions in international testing. The United States is currently twenty-fifth in life expectancy and eighteenth in diabetes rates but number one in obesity rates, and debt to our misfortune.[3] As a nation we should strive for better rankings in the STEM areas as well as in other domains of academic pursuit. While some would argue that STEM is not for everyone, our people are better than the numbers show.

The No Child Left Behind Act (NCLB) of 2001 was an attempt by the U.S. government's Bush administration to reverse the effects of low and uneven standards for achievement in the nation. The measure helped to rein in a lack of centralized control over education and institute standards across the United States. However, its work was not supported with resources in ways that would bring more people up to the levels of those standards developed.

NCLB and standards-based reform efforts were undertaken by all states with guidance from the federal government. It was an excellent first step in establishing standards for success and increasing accountability for student learning. Standards were enacted and assessed, and great care was taken by most organizations to ensure the tests fit the content to be measured. It was a well-intended move to bring the nation's STEM areas back to dominance, but the act had drawbacks.

Teachers and districts lacked resources to implement the changes. Teachers were strongly encouraged to increase students' scores. These pressures led instructors to "teach to the test" to gain the desired result. Making standardized tests a focus contributed to test anxiety (TA). An analysis of NAEP test results in twenty-five states from 1990 to 2003 showed several issues: 1) high-stakes testing will have a greater negative impact on minority students since states with larger minority groups have more of a focus on testing; 2) increases in standardized testing did not produce gains for fourth- and eighth-grade reading scores; and 3) high-stakes testing pressures were related to larger numbers of student dropouts due to TA.[4]

TA accompanies high-stakes testing, exacerbated by the NCLB reform movement, because many life decisions are based on a small set of standardized measures. TA is a difficult result of high-stakes testing of students. TA could be dealt with through developing instructional strategies to improve the symptoms early in a student's education. If K–6 education could intervene at the first signs of TA, students could manage it more easily. Perhaps a link of the tests to a more positive attitude about academic work could be strengthened. No one is going to love getting tested, but students can learn to love science by associating the tests with studying, learning, and a positive outcome. Improvements in science education, in accordance with recommendations held by this book, can foster better learning and success rates in STEM if they are linked to a clear alignment between standards and assessments.

Simply dismissing standards-based reform efforts is not a productive strategy for improving STEM education. Measured and sustained progress could have been implemented had NCLB had further support to include more science areas and funding for the specific needs of STEM education. Standardization in testing still lacks popular political support and serious financial extension because it is easier to criticize than to find methods of improvement.

Real successes stories, however, using standardized testing did result. Students in Massachusetts improved in their scores to match most of those in Europe, and other states improved significantly on many academic STEM measures. The MCAS (Massachusetts Comprehensive Assessment System) science exams are comparable to assessments in the best-performing nations. Professional involvement in developing standards and assessment for implementing NCLB as a science consultant and advisor since its inception was, personally, a hopeful endeavor. Standards-based reform, ranging from standard realignment to creating new tailored assessments, made testing quite professional. Curriculum was realigned to match the standards and assessments.

It is equally important to have the administrative support in each school to align instruction. Unfortunately, institutions had negative consequences for poor grades, and school district funding was often on the line. Administrations with more financial and structural support for addressing individual students' needs had the best outcomes. Poor schools had far worse results. As a result, NCLB was reduced to a game of schools competing for funding. Of course, one wonders, where is science education headed next in the absence of support for NCLB?

The current goal in standards-based school assessment is to have every person reach a level of science achievement under the No Child Left Behind Act of 2001 and Race to the Top of 2009. Both the Bush and Obama administrative plans recognize that all students can learn science to a level that will help them function better in life. Learning science and the scientific process is a noble and humane goal for improving people's lives.

Currently, however, there are no clear national initiatives for improving STEM education. Despite the STEM issues discussed, specified directives and a clear plan for moving forward are not being developed. In fact, the current U.S. administration and its Congress are set to cut funding to a number of STEM areas. Instead of plans, each state is attempting to obtain waivers to remove themselves from standards-based reform. These strategies may provide immediate budget relief, but at high costs to society at large. It is a chaos of change instead of systemwide agreed-upon improvements in STEM. Governmental strategies at this time offer little hope for STEM improvement.

Internationally, investment in STEM areas continues to grow. For example, China, Germany, and South Korea are diverting resources to STEM development. The United States is cutting back in these areas while increasing subsidies for consumption of foreign-made goods. Demand-side economics is being practiced, which rarely stimulates entrepreneurship. In the United States, monies are being given to the public to spend more on consumer products and small-ticket items. This is failing to create American jobs because these products are made at lower costs abroad.

Again, this is the opposite of what should be done to drive economic growth in STEM areas. New technologies and industries, as developed by STEM professionals, have long been the main driver in job creation and economic growth in this nation. The TV, radio, car, and computer were developed in this way to grow the economy.

As discussed earlier, today's infrastructure was built up by policies developed by generations before. Almost fifty years ago, the Eisenhower administration embarked on engineering projects and research to build the interstate highway system in the 1950s and early 1960s. It only takes a drive through the United States to recognize an infrastructure in decline. Rails and highways were implemented in the last half-century, but major changes in infrastructure did not materialize with the stimulus plans of the past few years. Tractor-trailers dominate the interstate roads, and the rail system is clearly inefficient, failing to meet transportation needs. These are only a few examples of how neglected STEM sections of the economy are leading to national weaknesses.

When one considers that a new innovation could have changed travel instead of leaving the roads to fall apart in neglect, one becomes disappointed. There have been innovators in the past decade, with creative businesspeople abundant. Steve Jobs and the Apple industry are a modern example, but where will STEM lead us in the near and distant future?

THE RISE OF THE REST

More goods and services are produced with fewer people as STEM products are implemented. For example, one tractor is more efficient than ten people working in a field. STEM does put many laborers out of work, which is a main reason to retrain them into a STEM career. The future economy requires more and more skills to be a successful employee. We have made many advances in these areas.

That said, other nations are catching up by playing the same game as we have been in the past century. Harvard historian Niall Ferguson, in his book *Civilization: The West and the Rest*, places this in historical context:

> For 500 years the West patented six killer applications that set it apart. The first to download them was Japan. Over the last century, one Asian country after another has downloaded these killer apps—competition, modern science, the rule of law and private property rights, modern medicine, the consumer society, and the work ethic. Those six things are the secret sauce of Western civilization.[5]

While our nation's dominance is driven by science, technology, and medicine, it is unclear whether we will remain on top, given the state of culture and STEM education.

IMPROVEMENTS TO SCIENCE EDUCATION ARE OUR LINK TO SUCCESS

Improvements in science education are likely to lead to better recruitment and retention of talented people in science. This will be accomplished by reducing student loss rates in college STEM courses; a change in the structure of the postsecondary STEM classroom; a shift in STEM instruction from simple knowledge transmission to actively and cooperatively engaging students; active student involvement in critical thinking and collaboration; higher student achievement; better science reasoning and thinking; more STEM cooperative learning strategies; and increased faculty involvement with students. These changes will improve student attitudes, achievement, and retention in STEM majors.

These changes will happen within the next decade because current research will continue to guide new instructors in those directions. It is obvious that merely presenting content in introductory college science courses without giving students opportunities for seeing how that knowledge fits within the whole curriculum is irresponsible. It does not further the kind of collateral, integrative learning advocated by national and state secondary standards in STEM areas. Although scientific factual knowledge is important, higher-order cognitive processes, which compare and connect the phenomena of a variety of academic areas are necessary in STEM.[6] Changes in instruction to reflect the themes found in this book should help attract students to STEM.

John Dewey (1859–1952), an early educational researcher, cited the importance of a principle of continuity between different subjects. This is reflected in my recommendations for content in postsecondary STEM courses in chapter 4, "Science for Every Person." A result should be that STEM students engage in collateral, synthetic learning. My recommendations will foster what Dewey considers "the most important attitude that can be formed[,] . . . that of a desire to go on learning."[7]

Student learning will be enhanced when we are able to help students see the relationships among the sciences and between science and mathematics, the humanities, social sciences, and the arts. Interdisciplinary approaches will guide STEM learning to areas of collateral interest. Developing curriculum by the organization of that content around themes, issues, or projects can enrich the students' view that sciences are not separable from other areas of life and can be reasoned about in a more holistic way.[8]

Traditional lectures, though, remain the most common form of instruction in introductory STEM courses.[9] I thus recommend a change from the lecture-oriented instruction. I emphasize the need for smaller class sizes (fewer than twenty-five students). The use of recitation sessions or extensive use of interactive learning in large classes may help, but it is difficult to foster higher-level learning with many students in a classroom.

Certain instructional strategies to enhance student learning in large lecture settings are beneficial, according to the research. For example, some advice for science instruction may be to use paradoxes and apparent contradictions to engage students, make connections with other courses and everyday phenomena, begin each class with something familiar to students, affect student motivation through your delivery (e.g., eye contact, enthusiasm), and ask *divergent* (having more than one answer) over *convergent questions* (having only one answer).[10] Each of these suggestions is also helpful in small classes. They emphasize student motivation and engagement—becoming active in the science classroom.

CONCLUSIONS

Change in the structure of the dominant culture is advocated to improve STEM achievements. A variety of cultural elements affect STEM education. There is an interaction of public opinion, science education, and government that may facilitate or act as a barrier to STEM achievement.

It would appear that an active national research program, perhaps tracking nonpersisting students in science, would be beneficial to improving retention. National standardization of assessment and content would be useful in improving scientific literacy. Through agreed-upon educational reforms, STEM alignment along the differing levels of study will be improved.

Action at the institutional level is the most important factor in improving STEM and the economy. Because a mismatch in student preparation between high school and college has been identified, STEM programs should safeguard against student losses due to preparation incongruities. These changes might include remediative courses or workshops to lessen possible academic deficiencies caused by preparation discrepancies, counseling services to help students cope with academic adjustments during their transition to college STEM, the recruitment of science faculty willing to participate in retention research, and the fostering of communication among institutions.

STEM fields have a great deal to offer; they are exciting and interesting. The challenge is to enhance their image to recruit and retain qualified students. That is a goal of this book, and it is hoped that the science, philosophy, and historical base, along with the tools developed to explore science, will encourage the reader to be a lifelong science learner. In this way, readers will have the desire to go on learning and sharing science to better understand and contribute to the world around them.

Notes

CHAPTER 1

1. Associated Press, IPSOS. "Poll: One-third in AP Poll Believe in Ghosts and UFOs, Half Accept ESP" (2007); www.ipsos-na.com/news-polls/pressrelease.aspx?id=3694. Accessed July 8, 2012.

2. Audrey Champagne, *The More Things Change . . .* (New York: Springer Science, in press).

3. H. J. Mackinder, "On the Scope and Methods of Geography," *Proceedings of the Royal Geographical Society and Monthly Record of Geography* 9, no. 3 (1887): 141–61. Quote is on page 154.

4. R. Kanai, B. Bahrami, Roylance, and G. Rees, "Abstract: Online social network site is reflected in human brain structure." *Proceedings of the Royal Society B* 279 (2011):1327.

5. Alice Calaprice, *The New Quotable Einstein* (Princeton, NJ: Princeton University Press, 2005), 238.

6. Dennis Sherman and Joyce Salisbury, *The West in the World* (New York: McGraw Hill Publishing, 2011).

7. Elaine Seymour and Nancy Hewitt, "Talking about Leaving: Factors Contributing to High Attrition Rates among Science, Math, and Engineering Undergraduate Engineering Majors, Final Report to the Alfred P. Sloan Foundation on an Ethnographic Inquiry at Seven Institutions" (Boulder: University of Colorado, 1994).

8. Alexander Astin, *Four Critical Years* (San Francisco: Jossey-Bass, 1977).

9. Alexander Astin and Helen S. Astin, *Undergraduate Science Education: The Impact of Different College Environments on the Educational Pipeline in the Sciences* (Los Angeles: Higher Education Research Institute, UCLA, 1993).

10. Elaine Seymour and Nancy Hewitt, *Talking about Leaving: Why Undergraduates Leave the Sciences* (Boulder: Westview Press, 1997).

11. R. Hilton and D. Lee, "Student Interest and Persistence in Science: Change in the Educational Pipeline in the Last Decade," *Journal of College Student Retention* 59 no. 5 (1988): 510–26.

12. Peter Daempfle, "Faculty Assumptions about the Student Characteristics Required for Success in Introductory College Biology" (PhD dissertation, Accession No. AAI9997550, University at Albany, 2000).

13. National Research Council, *National Science Education Standards* (Washington, DC: National Academy of Sciences Press, 1996).

14. Audrey Champagne and Leslie Hornig, *Science Teaching* (Washington, DC: American Association for the Advancement of Science, 1985).

15. Peter Daempfle, "An Analysis of the High Attrition Rates among First-Year College Science, Mathematics, and Engineering Majors," *Journal of College Student Retention* 5, no. 1 (2004): 37–52.

16. Daemple, "An Analysis of the High Attrition Rates."

CHAPTER 2

1. Peter Daempfle, "The Effects of Instructional Approaches on the Improvement of Reasoning in Introductory College Biology: A Quantitative Review of Research," *Bioscene: The Journal of College Biology Teaching* 32, no. 4 (2006): 22–32; J. B. Freeman, *Thinking Logically: Basic Concepts for Reasoning.* Englewood Cliffs, NJ: Prentice-Hall, 1988; T. Govier, *A Practical Study of Argument*, 3rd ed. (Belmont, CA: Wadsworth, 1992).

2. Theodore Schick Jr. and Lewis Vaughn, *How to Think about Weird Things: Critical Thinking for a New Age* (Mountain View, CA: Mayfield Publishing Company, 1995), 69–95.

3. Trudy Govier, *A Practical Study of Argument* (Florence, KY: Cengage/Wadsworth Publishing, 2004); Douglas Walton, *Argument Structure: A Pragmatic Theory* (Toronto, Canada: University of Toronto Press, 1996).

4. Gladys Leithauser and Marilyn Bell, *The World of Science: An Anthology for Writers* (New York: Holt, Rinehart and Winston, 1987).

5. R. Grant Steen, "Retractions in Medical Ethics: How Many Patients Are Put at Risk by Flawed Medical Research?" *Journal of Medical Ethics* 37 (2011): 688–92.

6. Jean Piaget, *Science of Education and the Psychology of the Child* (New York: Orion Press, 1970), 179.

7. R. Allen, "Intellectual Development and the Understanding of Science: Applications of William Perry's Theory to Science Teaching," *Journal of College Science Teaching* 12 (1981): 94–97.

8. B. Hofer and P. Pintrich, "The Development of Epistemological Theories: Beliefs about Knowledge and Knowing and Their Relation to Learning," *Review of Educational Research* 67, no. 1 (1997): 88–140; P. M. King and K. S. Kitchener, *Developing Reflective Judgment* (San Francisco: Jossey-Bass Publishers, 1994), 14–16.

9. Stathis Psillos, *Scientific Realism: How Science Tracks Truth* (London: Routledge, 1999).

CHAPTER 3

1. Center for Disease Control and Prevention, "Morbidity and Mortality Weekly Report (MMWR): Primary Amebic Meningoencephalitis—Arizona, Florida, and Texas, 2007," May 30, 2008. http://www.cdc.gov/mmwr/preview/mmwrhtml/mm5721a1.htm.

2. Richard Currey, National Institutes of Health, Office of Science Education, "Ulcers: The Culprit is H. Pylori!" http://science.education.nih.gov/home2.nsf/Educational +ResourcesResource+FormatsOnline+Resources+High+School/928BAB9A176A71B585256 CCD00634489.

3. Dennis Sherman and Joyce Salisbury, *The West in the World* (New York: McGraw Hill Publishing, 2011), 455.

4. G. Keppel, W. Saufley, and H. Tokunaga, *Introduction to Design and Analysis: A Student's Handbook, 2nd ed.* (New York: W. H. Freeman and Company, 1992).

5. Keppel, Saufley, and Tokunaga, *Introduction to Design and Analysis.*

6. Amy Kroenke, *Atmospheric Mercury Deposition to Sediments of New Jersey and Southern New York State: Interpretations from Dated Sediment Cores* (PhD dissertation, Troy, New York, Rensselaer Polytechnic Institute, 2003).

7. D. Rosenhan, "On Being Sane in Insane Places," *Science* 179, no. 4070 (1973): 250–58.

8. J. Lee, *The Scientific Endeavor: A Primer on Scientific Principles and Practice* (San Francisco: Addison Wesley Longman, 2000).

CHAPTER 4

1. American Association for the Advancement of Science, Project 2061, *Benchmarks for Science Literacy* (Washington, DC: American Association for the Advancement of Science, 1993).

2. National Research Council, *National Science Education Standards* (Washington, DC: National Academy of Sciences Press, 1996).

3. National Research Council, *National Science Education Standards.*

4. National Research Council, *National Science Education Standards.*

5. National Research Council, *National Science Education Standards.*

6. National Research Council, *National Science Education Standards.*

7. Vincent Tinto, *Leaving College: Rethinking the Causes and Cures of Student Attrition, 2nd ed.* (Chicago: University of Chicago Press, 1993).

8. Tinto, *Leaving College.*

9. C. Barrowman, "Improving Teaching and Learning Effectiveness by Defining Expectations," *New Directions for Higher Education* 96 (1996): 103–13.

10. Arnold Van Gennep, *The Rites of Passage* (Chicago: University of Chicago Press, 1960).

11. J. Chaskes, "The First-Year Student as Immigrant," *Journal of the Freshman Year Experience and Students in Transition* 8 (1996): 79–91.

12. Vincent Tinto, "Stages of Student Departure: Reflections on the Longitudinal Character of Student Leaving," *Journal of Higher Education* 59 (1989): 438–55.

13. Tinto, "Stages of Student Departure."

14. L. C. Attinasi Jr., "Getting In: Mexican Americans' Perceptions of University Attendance and the Implications for Freshman Year Persistence," *Journal of Higher Education* 60 (1989): 247–77.

15. E. Pascarella and P. Terenzini, *How College Affects Students* (San Francisco: Jossey-Bass, 1991).

16. Tinto, "Stages of Student Departure."

17. H. London, "Breaking Away: A Study of First-Generation College Students and Their Families," *American Journal of Sociology* 97 (1989): 144–70.

18. Rachelle Winkle-Wagner, "The Perpetual Homelessness of College Experiences: Tensions between Home and Campus for African-American Women," *Review of Higher Education* 33 (2009): 1–36.

19. Chaskes, "The First-Year Student as Immigrant."

20. I. Horowitz and W. Friedland, *The Knowledge Factory* (Carbondale, IL: Southern Illinois Press, 1972).

21. Chaskes, "The First-Year Student as Immigrant."

22. Tinto, "Stages of Student Departure."

23. Chaskes, "The First-Year Student as Immigrant."

24. Chaskes, "The First-Year Student as Immigrant."

25. Chaskes, "The First-Year Student as Immigrant."

26. Madge Lawrence Treeger and Karen Levin Coburn, *Letting Go* (New York: Harper Perrenial, 1997).

27. Peter Daempfle, "The Relationship between Stated and Actual Faculty Expectations for Introductory College Biology." Paper presented at the annual meeting of the New England Educational Research Organization, Portsmouth, 1999.

28. R. Gill, N. Glazer, and S. Thernstrom, *Our Changing Population* (Englewood Cliffs, NJ: Prentice Hall, 1992).

29. Peter Daempfle, "Faculty Assumptions about the Student Characteristics Required for Success in Introductory College Biology," *Bioscene: The Journal of College Biology Teaching* 28, no. 4 (2003): 19–33.

30. Daempfle, "Faculty Assumptions about the Student Characteristics."

31. Daempfle, "Faculty Assumptions about the Student Characteristics."

32. Daempfle, "Faculty Assumptions about the Student Characteristics."

33. California Higher Education System, *Statement on Preparation in Natural Science Expected of Entering College Freshmen: California Community Colleges* (Sacramento: California State University; Sacramento: Academic Senate, 1984). ERIC Document Reproduction Number ED 242375.

34. Daempfle, "Faculty Assumptions about the Student Characteristics."

35. Daempfle, "Faculty Assumptions about the Student Characteristics."

36. K. Tobin, D. Tippins, and A. Gallard, *Research on Instructional Strategies for Teaching Science. Handbook of Research on Science Teaching and Learning* (New York: MacMillan, 1994).

37. Daempfle, "Faculty Assumptions about the Student Characteristics."

38. T. Sizer, *Horace's School* (New York: Random House, 1995).

39. J. Dewey, *Experience and Education* (New York: Collier Macmillan Publishers, 1938).

40. Dewey, *Experience and Education.*

41. Dewey, *Experience and Education.*

CHAPTER 5

1. Brooke Noel Moore and Richard Parker, *Critical Thinking, 4th ed.* (Mountain View, CA: Mayfield Publishing Company, 1995), 4.

2. Wallace Stegner, *Beyond the Hundredth Meridian: John Wesley Powell and the Second Opening of the American West, Sentry Edition* (Boston: Houghton Mifflin Company, 1962), 214–15.

3. Daniel Chiras, *Environmental Science* (Burlington, MA: Jones and Bartlett Publishers, 2013).

4. Peter Daempfle, "The Effects of Instructional Approaches on the Improvement of Reasoning in Introductory College Biology: A Quantitative Review of Research," *Bioscene: The Journal of College Biology Teaching* 32, no. 4 (2006): 22–32.

5. Daempfle, "The Effects of Instructional Approaches."

6. National Research Council, *National Science Education Standards* (Washington, DC: National Academy of Sciences Press, 1997).

7. B. Hofer and P. Pintrich, "The Development of Epistemological Theories: Beliefs about Knowledge and Knowing and Their Relation to Learning," *Review of Educational Research* 67, no. 1 (1997): 88–140.

8. Daempfle, "The Effects of Instructional Approaches"; P. King and K. Kitchener, *Developing Reflective Judgment* (San Francisco: Jossey-Bass Publishers, 1994); W. Perry, *Forms of Intellectual and Ethical Development in the College Years: A Developmental Scheme* (New York: Holt, Rinehart, and Winston, 1970).

9. J. Piaget, *Science of Education and the Psychology of the Child* (New York: Orion Press, 1970).

10. R. Allen, "Intellectual Development and the Understanding of Science: Applications of William Perry's Theory to Science Teaching," *Journal of College Science Teaching* 12 (1981): 94–97.

11. King and Kitchener, *Developing Reflective Judgment.*

12. J. Barnard, "The Lecture-Demonstration vs. the Problem-Solving Method of Teaching a College Science Course," *Science Education* 26 (1942): 121–32.

13. A. Lawson and D. Snitgen, "Teaching Formal Reasoning in a College Biology Course for Preservice Teachers," *Journal of Research in Science Teaching* 19 (1982): 233–48; R. Tyser, and W. Cerbin, "Critical Thinking Exercises for Introductory Biology Courses," *Bioscience* 41 (1991): 41–46; M. Moll and R. Allen, "Developing Critical Thinking Skills in Biology," *Journal of College Science Teaching* 12 (1982): 95–98; D. Ebert-May, C. Brewer, and S. Allred, "Innovation in Large Lectures: Teaching for Active Learning," *Bioscience* 47, no. 9 (1997): 601–07.

14. G. Haukoos and J. Penick, "The Influence of Classroom Climate on Science Process and Content Achievement of Community College Students," *Journal of Research in Science Teaching* 20 (1983): 629–37.

15. Moore and Parker, *Critical Thinking,* 4th ed.

16. Jeffrey Lee, *The Scientific Endeavor: A Primer on Scientific Principles and Practice* (San Francisco: Addison Wesley Longman, 2000), 23–24.

17. Richard Paul, "Strategies: Thirty-Five Dimensions of Critical Thinking," in *Critical Thinking: What Every Person Needs to Survive in a Rapidly Changing World,* ed. A. J. A. Binker (Rohnert Park, CA: Foundation for Critical Thinking, 1992), 391–445.

18. Paul, "Strategies: Thirty-Five Dimensions of Critical Thinking."

CHAPTER 6

1. R. Kanai, B. Bahrami, R. Roylance, and G. Rees, "Online Social Network Size Is Reflected in Human Brain Structure," *Proceedings of the Royal Society B; Biological Sciences* 1–8 (2011). Accessed May 27, 2012. DOI: 10.1098/rspb.2011.1959.

2. Neil Postman, *Amusing Ourselves to Death: Public Discourse in the Age of Show Business* (New York: Penguin, 2005).

3. Postman, *Amusing Ourselves to Death.*

4. Postman, *Amusing Ourselves to Death.*

5. Postman, *Amusing Ourselves to Death.*

6. Postman, *Amusing Ourselves to Death.*

7. Postman, *Amusing Ourselves to Death.*

8. L. Parker, "Keyless Ignitions Eyed in Three Deaths," NBC Chicago.com, 2011. Last modified February 3, 2011. http://www.nbcchicago.com/news/business/parker-target-5-key-less-ignition-115241869.html.

9. Parker, "Keyless Ignitions Eyed in Three Deaths."

10. Parker, "Keyless Ignitions Eyed in Three Deaths."

CHAPTER 7

1. Immanel Velikovsky, *Worlds in Collision* (New York: Pocket Books, 1950).

2. P. B. Fontanarosa and G. B. Lundberg, "Alternative Medicine Meets Science," *Journal of the American Medical Association* 280, no. 18 (1998): 1618–19.

3. Jeffrey Lee, *The Scientific Endeavor: A Primer on Scientific Principles and Practices* (San Francisco: Addison Wesley Longman, Inc., 2000).

4. John D. McGervey, "A Statistical Test of Sun Sign Astrology," in *Paranormal Borderlands of Science*, ed. Kendrick Frazier (Buffalo: Prometheus Books, 1981), 235–40.

5. S. Carlson, "A Double-Blind Test of Astrology," *Nature* 318 (1985): 419–25.

6. Milton A. Rothman, *A Physicist's Guide to Skepticism* (Buffalo: Prometheus Books, 1988), 150.

7. Stewart Robb, *Nostradamus on Napoleon, Hitler and the Present Crisis* (New York: Charles Scribner's Sons, 1941), 171.

8. Lee, *The Scientific Endeavor.*

9. Lee, *The Scientific Endeavor.*

10. Lee, *The Scientific Endeavor.*

11. Lee, *The Scientific Endeavor.*

12. Lee, *The Scientific Endeavor.*

13. Terry Hamblin, "Selling America: The 'Voice of America' and United States International Radio Broadcasting to Western Europe during the Early Cold War, 1945–1954" (PhD dissertation, Stony Brook University, 2006).

14. Alice Calaprice, *The New Quotable Einstein* (Princeton, NJ: Princeton University Press, 2005), 238.

CHAPTER 8

1. Janice Thompson, *Science of Nutrition* (Boston: Benjamin Cummings, 2010).

2. Thompson, *Science of Nutrition.*

3. P. Havel, "Not All Sugars Are Equal, When It Comes to Weight Gain and Health," *Journal of Clinical Endocrinology & Metabolism* 89, no. 6 (2004).

4. G. Bray, S. Nielsen, and B. Popkin, "Consumption of High Fructose Corn Syrup in Beverages May Play a Role in the Epidemic of Obesity," *American Journal of Clinical Nutrition* 79 (2004): 537–43.

5. Connie Bennett and Stephen Sinatra, *Sugar Shock!: How Sweets and Simple Carbs Can Derail Your Life—and How You Can Get Back on Track* (New York: Berkley Books, 2007).

6. Harvey Diamond and Marilyn Diamond, *Fit for Life* (New York: Warner Books, 1985).

7. Stephen Barrett, "Weight Control: Facts, Fads, and Frauds," in *The Health Robbers: A Close Look at Quackery in America*, ed. Stephen Barrett and William T. Jarvis (Buffalo: Prometheus Books, 1993), 191–202.

8. Barrett, "Weight Control: Facts, Fads, and Frauds."

9. E. O. Smith, *When Culture and Biology Collide: Why We Are Stressed, Depressed, and Self-Obsessed* (New Brunswick, NJ: Rutgers University Press, 2002).

10. Smith, *When Culture and Biology Collide.*

11. Mark Cohen and George Armelagos, *Paleopathology at the Origins of Agriculture* (New York: New York Academic Press, 1984); George Armelagos, "Biocultural Aspects of Food Choice," in *Food and Evolution: Toward a Theory of Human Food Habits*, ed. M. Harris and Eric B. Ross (Philadelphia: Temple University Press, 1987).

12. "Deer Game Meat," FatSecret.com. http://www.fatsecret.com/calories-nutrition/usda/deer-game-meat (accessed June 24, 2012).

13. Cohen and Armelagos, *Paleopathology at the Origins of Agriculture.* Armelagos, "Biocultural Aspects of Food Choice.

14. Smith, *When Culture and Biology Collide.*

15. D. Bloch, "Salt and Gold," *Journal of Salt History* 7, no. 9 (1999); R. L. Hanneman, "Intersalt: Hypertension Rise with Age Revisited," *Journal of the American Medical Association* 312, no. 7041 (1996): 1283–84; discussion, 1284–87; T. A. Kotchen and R. M. Krauss, "Dietary Sodium and Blood Pressure," *Journal of the American Medical Association* 276, no. 18 (1996): 1468; discussion, 1469–70.

16. Bloch, "Salt and Gold"; Hanneman, "Intersalt: Hypertension Rise with Age Revisited"; Kotchen and Krauss, "Dietary Sodium and Blood Pressure."

17. Bloch, "Salt and Gold"; Hanneman, "Intersalt: Hypertension Rise with Age Revisited"; Kotchen and Krauss, "Dietary Sodium and Blood Pressure."

18. E. Ravussin et al., "The Effects of a Traditional Lifestyle on Obesity in Pima Indians," *Diabetes Care* 17, no. 9 (1994): 1067–74.

19. John La Puma and Rebecca Marx, *ChefMD's Big Book of Culinary Medicine: A Food Lover's Road Map to Losing Weight, Preventing Disease, and Getting Really Healthy* (New York: The Crowne Publishing Group, 2008).

20. Amy Kroenke, *Atmospheric Mercury Deposition to Sediments of New Jersey and Southern New York State: Interpretations from Dated Sediment Cores* (PhD dissertation, Troy, New York, Rensselaer Polytechnic Institute, 2003).

21. Michael Behe, *Darwin's Black Box: The Biochemical Challenge to Evolution* (New York: Free Press, 2006).

22. Russell Doolittle, "A Delicate Balance," *Boston Review*, February–March 1997. http://bostonreview.net/BR22.1/doolittle.html.

23. J. D. Borucinska, J. C. Harshbarger, and T. Bogicevic, "Hepatic Cholangiocarcinoma and Testicular Mesothelioma in Wild-Caught Blue Shark, Prionace glauca (L.)," *Journal of Fish Diseases* 26, no. 1 (2003): 43–49.

24. K. Kruszelnicki, "Ostrich Head in Sand," *ABC Science: In Depth*. Australian Broadcast Company, 2006. http://www.abc.net.au/science/articles/2006/11/02/1777947.htm.

25. K. Grammer and R. Thornhill, "Human (*Homo sapiens*) Facial Attractiveness and Sexual Selection: The Role of Symmetry and Averageness," *Journal of Comparative Psychology* 108, no. 3 (1997): 233–42.

26. Smith, *When Culture and Biology Collide*.

27. Smith, *When Culture and Biology Collide*.

28. Ronald Nowak and Ernest Walker, *Walker's Mammals of the World, vol. 1* (Baltimore, MD: Johns Hopkins University Press, 1991).

29. Smith, *When Culture and Biology Collide*.

30. Richard Dawkins, *The Selfish Gene* (New York: Oxford University Press, 1976).

31. Konrad Lorenz, *On Aggression* (Orlando, FL: Mariner Books, 1974).

32. Terence Bazzett, *An Introduction to Behavior Genetics* (Sunderland, MA: Sinauer and Associates, 2008).

33. M. J. Lyons, K. C. Koenen, F. Buchting, and J. M. Meyer et al., "A Twin Study of Sexual Behavior in Men," *Archives of Sexual Behavior* 33, no. 2 (2004): 129–36; F. L. Whitam, M. Diamond, and J. Martin, "Homosexual Orientation in Twins: A Report on 61 Pairs and Three Triplet Sets," *Archives of Sexual Behavior* 22, no. 3 (1993): 187–206.

34. Bazzett, *An Introduction to Behavior Genetics*.

35. Bazzett, *An Introduction to Behavior Genetics*.

36. A. Helgason, "An Association between the Kinship and Fertility of Human Couples," *Science, The Science Creative Quarterly* 391 (2003): 813–16.

37. Smith, *When Culture and Biology Collide*.

38. Smith, *When Culture and Biology Collide*.

39. Helena Curtis, *Biology* (New York: Worth Publishers, 1986).

40. Marieb, Elaine, and Katja Hoehn, *Human Anatomy and Physiology*, 8th ed. San Francisco: Benjamin Cummings, 2010.

41. M. Odeh, H. Bassan, and A. Oliven, "Termination of Intractable Hiccups with Digital Rectal Massage," *Journal of Internal Medicine* 227, no. 2 (February 1990): 145–46.

42. Elizabeth B. Claus, Lisa Calvocoressi, Melissa Bondy, Joellen Schildkraut, Joseph Wiemels, and Margaret Wrensch, "Dental X-Rays and Risk of Meningioma," *Cancer*, April 12, 2012.

43. Smith, *When Culture and Biology Collide*.

44. C. D. Klose, "Human-Triggered Earthquakes and Their Impacts on Human Security, Achieving Environmental Security: Ecosystem Services and Human Welfare," in "NATO Science for Peace and Security Series-E: Human and Societal Dynamics," vol. 69 (2010): 13–19, ed. P. H. Liotta et al.

CHAPTER 9

1. C. Zahn and M. Miller, "Excess Length of Stay, Charges, and Mortality Attributable to Medical Injuries as a Result of Hospitalization," *American Medical Association* 14 (2005): 1868–74.

2. Robert Bell, *Impure Science: Fraud, Compromise, and Political Influence in Scientific Research* (New York: John Wiley & Sons, 1992).

3. Jeffrey Lee, *The Scientific Endeavor: A Primer on Scientific Principles and Practice* (San Francisco: Addison Wesley Longman, 2000).

4. Lee, *The Scientific Endeavor.*

5. Committee on Science, Engineering and Public Policy (U.S.) Panel on Scientific Responsibility and the Conduct of Research, *Responsible Science: Ensuring the Integrity of the Research Process, vol. I* (Washington, DC: National Academy Press, 1992), 5–6.

6. Lee, *The Scientific Endeavor.*

7. Francis L. Macrina, *Scientific Integrity: An Introductory Text with Cases* (Washington, DC: ASM Press, 1995), 47.

8. Max Otto, *Natural Laws and Human Hopes* (New York: Holt & Company, 1926).

9. James Trefil and Robert Hazen, *Science Matters: Achieving Scientific Literacy* (Harpswell, ME: Anchor Publishers, 2009).

10. O. LaFarge, "Scientists Are Lonely Men," *Harpers Magazine*, November 1942, 652–59; quote is on page 657.

11. Lee, *The Scientific Endeavor.*

12. Lee, *The Scientific Endeavor.*

13. Thomas Hager, *Force of Nature: The Life of Linus Pauling* (New York: Simon & Schuster, 1995), 573–627.

14. Carl Sagan, *The Demon-Haunted World: Science as a Candle in the Dark* (New York: Random House, 1995), 28.

15. Lee, *The Scientific Endeavor.*

16. Diana Crane, *Invisible Colleges: Diffusion of Knowledge in Scientific Communities* (Chicago: University of Chicago Press, 1972).

17. Lee, *The Scientific Endeavor.*

18. Brian VanDeMark, *Pandora's Keepers: Nine Men and the Atomic Bomb* (Boston: Back Bay Books, 2005).

CHAPTER 10

1. Jacques Nicolle, *Louis Pasteur: The Story of His Major Discoveries* (New York: Basic Books, Inc., 1961).

2. Theodore Zeldin, *An Intimate History of Humanity* (New York: HarperCollins, 1994).

3. Julian Vincent, *Structural Biomaterials* (Princeton, NJ: Princeton University Press, 1990).

4. Dennis Sherman and Joyce Salisbury, *The West in the World* (New York: McGraw Hill Publishing, 2011).

5. Sherman and Salisbury, *The West in the World.*

6. Sherman and Salisbury, *The West in the World.*

7. John McKay, Bennett Hill, John Buckler, Clare Haru Crowston, Merry Wiesner-Hanks, and Joe Perry, *A History of Western Society, 10th ed.* (Boston: Bedford/St. Martins, 2011).

8. Carl Sagan, *The Dragons of Eden: Speculation on the Evolution of Human Intelligence* (New York: Ballantine Books, 1977).

9. Sherman and Salisbury, *The West in the World.*

10. Peter Daempfle, "The Effects of Instructional Approaches on the Improvement of Reasoning in Introductory College Biology: A Quantitative Review of Research," *Bioscene: The Journal of College Biology Teaching* 32, no. 4 (2006), 22–32.

11. Martinez Ricardo, Statement of the Honorable Ricardo Martinez, M.D. administrator, National Highway Traffic Safety Commission, U.S. House of Representatives, Subcommittee on Surface Transportation, Washington, DC, 1997.

CHAPTER 11

1. D.W. Chambers, "Stereotypic Images of the Scientist: The Draw a Scientist Test." *Science Education*, 67, no. 2 (1983): 255–65.

2. "Stem Competency a Foundational Skill: Job Expert says." *US News and World Report*, October 20, 2011. Retrieved from USnews.com/news/blogs/stem-education/2011/10/20/stem-competency-a-foundational-skill-jobs-expert-says

3. Steve H. Murdock and David Swanson, *Applied Demography in the 21st Century* (New York: Springer/Verlag, 2008).

4. J. P. Michau, *European Unemployment: How Significan was a Declining Work Ethic. Centrepiece* (Autumn 2009). World Values Survey: http://cep.lse.ac.uk/pubs/download/cp294.pdf

5. Peter Daempfle, "An Analysis of the High Attrition Rates among First-Year College Science, Mathematics, and Engineering Majors," *Journal of College Student Retention* 5, no. 1 (2004): 37–52.

6. Alexander Astin and Helen Astin Sol, *Undergraduate Science Education: The Impact of Different College Environments and the Educational Pipeline in the Colleges* (Los Angeles: Higher Education Research Institute, UCLA, 1993).

7. Sylvai Hurtado, John Pryor et al. HERI Research Brief, *Degrees of Success: Bachelor's Degree Completion Rates among Initial STEM Majors* (Los Angeles: Higher Education Research Institute at UCLA, 2010).

8. Center for Institutional Data Exchange and Analysis, *1999–2000 SMET Retention Report* (Norman, OK: University of Oklahoma, 2000).

9. L. S. Sax, Hurtado, J. A. Lindholm, A. W. Astin, W. S. Korn, and K. M. Mahoney, *The American Freshman: National Norms for the Fall 2004* (Los Angeles: Higher Education Research Institute, UCLA, 2005).

10. Astin and Astin. *Undergraduate Science Education.*

11. Elaine Seymour and Nancy Hewitt, *Talking about Leaving: Why Undergraduates Leave the Science.* (Boulder: Westview Press, 1997).

12. National Research Council. *National Science Education Standards* (Washington, DC: National Academy of Sciences Press, 1996).

13. Daempfle, "An Analysis of the High Attrition Rates.

14. Seymour and Hewitt. *Talking about Leaving*, 1997.

15. Peter Daempfle, "An Analysis of the High Attrition Rates Among First-Year College Science, Mathematics, and Engineering Majors," 2004.

16. Daempfle, "An Analysis of the High Attrition Rates.

17. Seymour and Hewitt. *Talking about Leaving.*

18. Peter Daempfle, "Faculty Assumptions about the Student Characteristics Required for Success in Introductory College Biology," *Bioscene: The Journal of College Biology Teaching*, 28, no. 4 (2003): 19–33.

19. Daempfle, "Faculty Assumptions about the Student Characteristics Required for Success in Introductory College Biology."

20. Daempfle, "Faculty Assumptions about the Student Characteristics Required for Success in Introductory College Biology."

21. Daempfle, "Faculty Assumptions about the Student Characteristics Required for Success in Introductory College Biology."

22. C. Strenta, R. Elliott, A. Russell, M. Matier, and J. Scott. "Choosing and Leaving Science in Highly Selective Institutions." *Research in Higher Education,* 35 (1994): 513–37.

23. Seymour and Hewitt, *Talking about Leaving.*

24. Seymour and Hewitt. *Talking about Leaving.*

25. Judy McShannon, and Roberta Derlin. "Retaining Minority Women Engineering Students: How Faculty Development and Research Can Foster Student Success." Paper presented at the New Mexico Higher Education Assessment Conference, Las Cruces, New Mexico, 2000.

26. McShannon and Derlin, "Retaining Minority Wome Engineering Students."

27. Daempfle, "Faculty Assumptions about the Student Characteristics Required for Success in Introductory College Biology."

28. Daempfle, "An Analysis of the High Attrition Rates."

29. Daempfle, "An Analysis of the High Attrition Rates."

30. Daempfle, "An Analysis of the High Attrition Rates."

31. Daempfle, "An Analysis of the High Attrition Rates."

32. Schommer, M. "Comparisons of Beliefs about the Nature of Knowledge and Learning among Postsecondary Students." *Journal of Research in Higher Education*, 34 (1993): 355–68.

33. Schommer, "Comparisons of Belief."

CHAPTER 12

1. Peter Daempfle, "An Analysis of the High Attrition Rates among First-Year College Science, Mathematics, and Engineering Majors," *Journal of College Student Retention* 5, no. 1 (2004): 37–52.

2. J. Koebler, "Demand, Pay for STEM Skills Skyrocket," *US News and World Report*, October 20, 2011. http://www.usnews.com/news/blogs/stem-education/2011/10/20/stem-competency-a-foundational-skill-jobs-expert-says.

3. Fareed Zakaria, "Are America's Best Days Behind Us?" *Time*, March 31, 2011. http://fareedzakaria.com/2011/03/03/are-americas-best-days-behind-us/.

4. M. Hurley and F. Padro, "Test Anxiety and High Stakes Testing: Pervasive, Pernicious, Punitive and Policy-Driven," *International Journal of Learning* 13 (2006): 163–70.

5. Zakaria, "Are America's Best Days Behind Us?"

6. John Dewey, *Experience and Education* (New York: Collier Macmillan Publishers, 1938).

7. John Dewey, *Experience and Education* (New York: Collier MacMillan Publishers, 1938).

8. National Academies Press, *Science Teaching Reconsidered: A Handbook* (Washington, DC: Committee on Undergraduate Science Education, 1997).

9. Daempfle, "An Analysis of the High Attrition Rates."

10. National Academies Press, *Science Teaching Reconsidered.*

References

Allen, R. "Intellectual Development and the Understanding of Science: Applications of William Perry's Theory to Science Teaching." *Journal of College Science Teaching* 12 (1981): 94–97.

American Association for the Advancement of Science. Project 2061. *Benchmarks for Science Literacy.* Washington, DC: American Association for the Advancement of Science, 1993.

Armelagos, George. "Biocultural Aspects of Food Choice." In *Food and Evolution: Toward a Theory of Human Food Habits,* ed. M. Harris and Eric B. Ross. Philadelphia, PA: Temple University Press, 1987.

Astin, Alexander. *Four Critical Years.* San Francisco: Jossey-Bass, 1977.

Astin, Alexander, and Helen S. Astin. *Undergraduate Science Education: The Impact of Different College Environments on the Educational Pipeline in the Sciences.* Los Angeles: Higher Education Research Institute, UCLA, 1993.

Attinasi, L. C., Jr. "Getting In: Mexican Americans' Perceptions of University Attendance and the Implications for Freshman Year Persistence." *Journal of Higher Education* 60 (1989): 247–77.

Barnard, J. "The Lecture-Demonstration vs. the Problem-Solving Method of Teaching a College Science Course." *Science Education* 26 (1942): 121–32.

Barrett, Stephen. "Weight Control: Facts, Fads, and Frauds." In *The Health Robbers: A Close Look at Quackery in America,* ed. Stephen Barrett and William T. Jarvis. Buffalo: Prometheus Books, 1993, 191–202.

Barrowman, C. "Improving Teaching and Learning Effectiveness by Defining Expectations." *New Directions for Higher Education* 96 (1996): 103–13.

Bazzett, Terence. *An Introduction to Behavior Genetics.* Sunderland, MA: Sinauer and Associates, 2008.

Behe, Michael. *Darwin's Black Box: The Biochemical Challenge to Evolution.* New York: Free Press, 2006.

Bell, Robert. *Impure Science: Fraud, Compromise, and Political Influence in Scientific Research.* New York: John Wiley & Sons, 1992.

Bennett, Connie, and Stephen Sinatra. *Sugar Shock!: How Sweets and Simple Carbs Can Derail Your Life—and How You Can Get Back on Track*. New York: Berkley Books, 2007.

Bloch, D. "Salt and Gold." *Journal of Salt History* 7, no. 9 (1999): 1283–84; discussion, 1284–87.

Borucinska, J. D., J. C. Harshbarger, and T. Bogicevic. "Hepatic Cholangiocarcinoma and Testicular Mesothelioma in Wild-Caught Blue Shark, Prionace glauca (L.)." *Journal of Fish Diseases* 26, no. 1 (2003): 43–49.

Bray, G., S. Nielsen, and B. Popkin. "Consumption of High Fructose Corn Syrup in Beverages May Play a Role in the Epidemic of Obesity." *American Journal of Clinical Nutrition* 79 (2004): 537–43.

Calaprice, Alice. *The New Quotable Einstein*. Princeton, NJ: Princeton University Press, 2005, 238.

California Higher Education System. *Statement on Preparation in Natural Science Expected of Entering College Freshmen: California Community Colleges*. Sacramento: California State University; Sacramento: Academic Senate, 1984. ERIC Document Reproduction Number ED 242375.

Carlson, S. "A Double-Blind Test of Astrology." *Nature* 318 (1985): 419–25.

Center for Disease Control and Prevention. "Morbidity and Mortality Weekly Report (MMWR): Primary Amebic Meningoencephalitis—Arizona, Florida, and Texas, 2007," May 30, 2008. http://www.cdc.gov/mmwr/preview/mmwrhtml/mm5721a1.htm.

Center for Institutional Data Exchange and Analysis. *1999–2000 SMET Retention Report*. Norman, OK: University of Oklahoma, 2000.

Chambers, D. W. "Stereotypic Images of the Scientist: The Draw-a-Scientist Test." *Science Education* 67, no. 2 (1983): 255–65.

Champagne, Audrey. *The More Things Change . . .* New York: Springer Science, in press.

Champagne, Audrey, and Leslie Hornig. *Science Teaching*. Washington, DC: American Association for the Advancement of Science, 1985.

Chaskes, J. "The First-Year Student as Immigrant." *Journal of the Freshman Year Experience and Students in Transition* 8 (1996): 79–91.

Chiras, Daniel. *Environmental Science*. Burlington, MA: Jones and Bartlett Publishers, 2013.

Claus, E. B., L. Calvocoressi, M. Bondy, J. Schildkraut, J. Wiemels, and M. Wrensch. "Dental X-Rays and Risk of Meningioma." *Cancer*, April 12, 2012.

Cohen, Mark, and George Armelagos. *Paleopathology at the Origins of Agriculture*. New York: New York Academic Press, 1984.

Committee on Science, Engineering and Public Policy (U.S.) Panel on Scientific Responsibility and the Conduct of Research. *Responsible Science: Ensuring the Integrity of the Research Process, vol. I*. Washington, DC: National Academy Press, 1992, 5–6.

Crane, Diana. *Invisible Colleges: Diffusion of Knowledge in Scientific Communities*. Chicago: University of Chicago Press, 1972.

Currey, Richard. National Institutes of Health, Office of Science Education. "Ulcers: The Culprit is H. Pylori!" http://science.education.nih.gov/home2.nsf/Educational+ResourcesResource+FormatsOnline+Resources+High+School/928BAB9A176A71B585256CCD00634489.

Curtis, Helena. *Biology*. New York: Worth Publishers, 1986.

Daempfle, Peter. "The Relationship between Stated and Actual Faculty Expectations for Introductory College Biology." Paper presented at the annual meeting of the New England Educational Research Organization, Portsmouth, 1999.

———. "Faculty Assumptions about the Student Characteristics Required for Success in Introductory College Biology." PhD dissertation, Accession No. AAI9997550, University at Albany, 2000.

———. "Faculty Assumptions about the Student Characteristics Required for Success in Introductory College Biology." *Bioscene: The Journal of College Biology Teaching* 28, no. 4 (2003): 19–33.

———. "An Analysis of the High Attrition Rates among First-Year College Science, Mathematics, and Engineering Majors." *Journal of College Student Retention* 5, no. 1 (2004): 37–52.

———. "The Effects of Instructional Approaches on the Improvement of Reasoning in Introductory College Biology: A Quantitative Review of Research." *Bioscene: The Journal of College Biology Teaching* 32, no. 4 (2006): 22–32.

Dawkins, Richard. *The Selfish Gene.* New York: Oxford University Press, 1976.

"Deer Game Meat." FatSecret.com. http://www.fatsecret.com/calories-nutrition/usda/deer-game-meat (accessed June 24, 2012).

Dewey, John. *Experience and Education.* New York: Collier Macmillan Publishers, 1938.

Diamond, Harvey, and Marilyn Diamond. *Fit for Life.* New York: Warner Books, 1985.

Doolittle, Russell. "A Delicate Balance," *Boston Review,* February–March 1997. http://bostonreview.net/BR22.1/doolittle.html.

Ebert-May, D., C. Brewer, and S. Allred. "Innovation in Large Lectures: Teaching for Active Learning." *Bioscience* 47, no. 9 (1997): 601–07.

Fontanarosa, P. B., and G. B. Lundberg. "Alternative Medicine Meets Science." *Journal of the American Medical Association* 280, no. 18 (1998): 1618–19.

Freeman, J. B. *Thinking Logically: Basic Concepts for Reasoning.* Englewood Cliffs, NJ: Prentice-Hall, 1988.

Gill, R., N. Glazer, and S. Thernstrom. *Our Changing Population.* Englewood Cliffs, NJ: Prentice Hall, 1992.

Govier, T. *A Practical Study of Argument,* 3rd ed. Belmont, CA: Wadsworth, 1992.

Govier, Trudy. *A Practical Study of Argument.* Florence, KY: Cengage/Wadsworth Publishing, 2004.

Grammer, K., and R. Thornhill. "Human (*Homo sapiens*) Facial Attractiveness and Sexual Selection: The Role of Symmetry and Averageness." *Journal of Comparative Psychology* 108, no. 3 (1997): 233–42.

Hager, Thomas. *Force of Nature: The Life of Linus Pauling.* New York: Simon & Schuster, 1995.

Hamblin, Terry. "Selling America: The 'Voice of America' and United States International Radio Broadcasting to Western Europe during the Early Cold War, 1945–1954." PhD dissertation, Stony Brook University, 2006.

Hanneman, R. L. "Intersalt: Hypertension Rise with Age Revisited." *Journal of the American Medical Association* 312, no. 7041 (1996): 1283–84.

Hager, Thomas. *Force of Natures: The Life of Linus Pauling.* New York: Simon & Schuster, 1995.

Haukoos, G., and J. Penick. "The Influence of Classroom Climate on Science Process and Content Achievement of Community College Students." *Journal of Research in Science Teaching* 20 (1983): 629–37.

H., Peter. "Not All Sugars Are Equal, When It Comes to Weight Gain and Health." *Journal of Clinical Endocrinology & Metabolism* 89, no. 6 (2004).

Helgason, A. "An Association between the Kinship and Fertility of Human Couples." *Science, The Science Creative Quarterly* 391 (2003): 813–16.

Hilton, R., and D. Lee. "Student Interest and Persistence in Science: Change in the Educational Pipeline in the Last Decade." *Journal of College Student Retention* 59, no. 5 (1988): 510–26.

Hofer, B., and P. Pintrich. "The Development of Epistemological Theories: Beliefs about Knowledge and Knowing and Their Relation to Learning." *Review of Educational Research* 67, no. 1 (1997): 88–140.

Horowitz, Irving Louis, and William Friedland. *The Knowledge Factory.* Carbondale, IL: Southern Illinois Press, 1972.

Hurley, M., and F. Padro. "Test Anxiety and High Stakes Testing: Pervasive, Pernicious, Punitive and Policy-Driven." *International Journal of Learning* 13 (2006): 163–70.

Hurtado, Sylvia, John Pryor et al. HERI Research Brief, *Degrees of Success: Bachelor's Degree Completion Rates among Initial STEM Majors.* Los Angeles: Higher Education Research Institute at UCLA, 2010.

Kanai, R., B. Bahrami, R. Roylance, and G. Rees. "Online Social Network Size Is Reflected in Human Brain Structure." *Proceedings of the Royal Society B: Biological Sciences* 1–8 (2011). Accessed May 27, 2012. DOI: 10.1098/rspb.2011.1959.

Keppel, Geoffrey, William H. Saufley, and Howard Tokunaga. *Introduction to Design and Analysis: A Student's Handbook, 2nd ed.* New York: W. H. Freeman and Company, 1992.

King, Patricia M., and Karen Strohm Kitchener. *Developing Reflective Judgment.* San Francisco: Jossey-Bass Publishers, 1994, 14–16.

Klose, C. D. "Human-Triggered Earthquakes and Their Impacts on Human Security, Achieving Environmental Security: Ecosystem Services and Human Welfare." In "NATO Science for Peace and Security Series-E: Human and Societal Dynamics," vol. 69(2010): 13–19, ed. P. H. Liotta et al.

Koebler, J. "Demand, Pay for STEM Skills Skyrocket," *US News and World Report,* October 20, 2011. http://www.usnews.com/news/blogs/stem-education/2011/10/20/stem-competency-a-foundational-skill-jobs-expert-says.

Kotchen, T. A., and R. M. Krauss. "Dietary Sodium and Blood Pressure." *Journal of the American Medical Association* 276, no. 18 (1996): 1468; discussion, 1469–70.

Kroenke, Amy. *Atmospheric Mercury Deposition to Sediments of New Jersey and Southern New York State: Interpretations from Dated Sediment Cores.* PhD dissertation, Troy, New York: Rensselaer Polytechnic Institute, 2003.

Kruszelnicki, K. "Ostrich Head in Sand." *ABC Science: In Depth.* Australian Broadcast Company, 2006. http://www.abc.net.au/science/articles/2006/11/02/1777947.htm.

LaFarge, O. "Scientists Are Lonely Men." *Harpers Magazine,* November 1942, 652–59.

La Puma, John, and Rebecca Marx. *ChefMD's Big Book of Culinary Medicine: A Food Lover's Road Map to Losing Weight, Preventing Disease, and Getting Really Healthy.* New York: The Crowne Publishing Group, 2008.

Lawson, A., and D. Snitgen. "Teaching Formal Reasoning in a College Biology Course for Preservice Teachers." *Journal of Research in Science Teaching* 19 (1982): 233–48.

Lee, Jeffrey. *The Scientific Endeavor: A Primer on Scientific Principles and Practice.* San Francisco: Addison Wesley Longman, 2000.

Leithauser, Gladys, and Marilyn Bell. *The World of Science: An Anthology for Writers.* New York: Holt, Rinehart and Winston, 1987.

London, H. "Breaking Away: A Study of First-Generation College Students and Their Families." *American Journal of Sociology* 97 (1989): 144–70.

Lorenz, Konrad. *On Aggression.* Orlando, FL: Mariner Books, 1974.

Lyons, M. J., K. C. Koenen, F. Buchting, and J. M. Meyer et al. "A Twin Study of Sexual Behavior in Men." *Archives of Sexual Behavior* 33, no. 2 (2004): 129–36.

Mackinder, H. J. "On the Scope and Methods of Geography." *Proceedings of the Royal Geographical Society and Monthly Record of Geography* 9, no. 3 (1887): 141–61.

Macrina, Francis L. *Scientific Integrity: An Introductory Text with Cases.* Washington, DC: ASM Press, 1995, 47.

McGervey, John D. "A Statistical Test of Sun Sign Astrology." In *Paranormal Borderlands of Science*, ed. Kendrick Frazier. Buffalo: Prometheus Books, 1981, 235–40.

McKay, John, Bennett Hill, John Buckler, Clare Haru Crowston, Merry Wiesner-Hanks, and Joe Perry. *A History of Western Society*, 10th ed. Boston: Bedford/St. Martins, 2011.

McShannon, Judy, and Roberta Derlin. "Retaining Minority Women Engineering Students: How Faculty Development and Research Can Foster Student Success." Paper presented at the New Mexico Higher Education Assessment Conference, Las Cruces, New Mexico, 2000.

Michau, J.-B. *European Unemployment: How Significant Was a Declining Work Ethic. CentrePiece* (Autumn 2009). http://cep.lse.ac.uk/pubs/download/cp294.pdf.

Moll, M., and R. Allen. "Developing Critical Thinking Skills in Biology." *Journal of College Science Teaching* 12 (1982): 95–98.

Moore, Brooke Noel, and Richard Parker. *Critical Thinking, 4th ed.* Mountain View, CA: Mayfield Publishing Company, 1995, 4.

Murdock, Steve H., and David Swanson. *Applied Demography in the 21st Century.* New York: Springer/Verlag, 2008.

National Academies Press. *Science Teaching Reconsidered: A Handbook.* Washington, DC: Committee on Undergraduate Science Education, 1997.

National Research Council. *National Science Education Standards.* Washington, DC: National Academy of Sciences Press, 1996.

———. *National Science Education Standards.* Washington, DC: National Academy of Sciences Press, 1997.

Nicolle, Jacques. *Louis Pasteur: The Story of His Major Discoveries.* New York: Basic Books, Inc., 1961.

Nowak, Ronald, and Ernest Walker. *Walker's Mammals of the World, vol. 1.* Baltimore, MD: Johns Hopkins University Press, 1991.

Odeh, M., H. Bassan, and A. Oliven. "Termination of Intractable Hiccups with Digital Rectal Massage." *Journal of Internal Medicine* 227, no. 2 (February 1990): 145–46.

Otto, Max. *Natural Laws and Human Hopes.* New York: Holt & Company, 1926.

Parker, L. 2011. "Keyless Ignitions Eyed in Three Deaths." NBC Chicago.com, 2011. Last modified February 3, 2011. http://www.nbcchicago.com/news/business/parker-target-5-keyless-ignition-115241869.html.

Pascarella, Ernest, and Patricia Terenzini. *How College Affects Students.* San Francisco: Jossey-Bass, 1991.

Paul, Richard. "Strategies: Thirty-Five Dimensions of Critical Thinking. In *Critical Thinking: What Every Person Needs to Survive in a Rapidly Changing World*, ed. A. J. A. Binker. Rohnert Park, CA: Foundation for Critical Thinking, 1992, 391–445.

Perry, William. *Forms of Intellectual and Ethical Development in the College Years: A Developmental Scheme.* New York: Holt, Rinehart, and Winston, 1970.

Piaget, Jean. *Science of Education and the Psychology of the Child.* New York: Orion Press, 1970.

Postman, Neil. Amusing *Ourselves to Death: Public Discourse in the Age of Show Business*. New York: Penguin, 2005.

Psillos, Stathis. *Scientific Realism: How Science Tracks Truth*. London: Routledge, 1999.

Ravussin, E., et al. "The Effects of a Traditional Lifestyle on Obesity in Pima Indians." *Diabetes Care* 17, no. 9 (1994): 1067–74.

Ricardo, Martinez. Statement of the Honorable Ricardo Martinez, M.D. administrator, National Highway Traffic Safety Commission, U.S. House of Representatives, Subcommittee on Surface Transportation, Washington, DC, 1997.

Robb, Stewart. *Nostradamus on Napoleon, Hitler and the Present Crisis*. New York: Charles Scribner's Sons, 1941, 171.

Rosenhan, D. "On Being Sane in Insane Places." *Science* 179, no. 4070 (1973): 250–58.

Rothman, Milton A. *A Physicist's Guide to Skepticism*. Buffalo: Prometheus Books, 1988, 150.

Sagan, Carl. "A Cosmic Calendar," *Natural History*, December 1975, 70–73.

———. *The Demon-Haunted World: Science as a Candle in the Dark*. New York: Random House, 1995, 28.

Sax, L. J., S. Hurtado, J. A. Lindholm, A. W. Astin, W. S. Korn, and K. M. Mahoney. *The American Freshman: National Norms for the Fall 2004*. Los Angeles: Higher Education Research Institute, UCLA, 2005.

Schick, Theodore Jr., and Lewis Vaughn. *How to Think about Weird Things: Critical Thinking for a New Age*. Mountain View, CA: Mayfield Publishing Company, 1995, 69–95.

Seymour, Elaine, and Nancy Hewitt. "Talking about Leaving: Factors Contributing to High Attrition Rates among Science, Math, and Engineering Undergraduate Engineering Majors, Final Report to the Alfred P. Sloan Foundation on an Ethnographic Inquiry at Seven Institutions." Boulder: University of Colorado, 1994.

———. *Talking about Leaving: Why Undergraduates Leave the Sciences*. Boulder: Westview Press, 1997.

Sherman, Dennis, and Joyce Salisbury. *The West in the World*. New York: McGraw Hill Publishing, 2011.

Sizer, Theodore R. *Horace's School*. New York: Random House, 1995.

Smith, E. O. *When Culture and Biology Collide: Why We Are Stressed, Depressed, and Self-Obsessed*. New Brunswick, NJ: Rutgers University Press, 2002.

Steen, R. Grant. "Retractions in Medical Ethics: How Many Patients Are Put at Risk by Flawed Medical Research?" *Journal of Medical Ethics* 37 (2011): 688–92.

Stegner, Wallace. *Beyond the Hundredth Meridian: John Wesley Powell and the Second Opening of the American West, Sentry Edition*. Boston: Houghton Mifflin Company, 1962, 214–15.

Strenta, C., R. Elliott, R. Adair, M. Matier, and J. Scott. "Choosing and Leaving Science in Highly Selective Institutions." *Research in Higher Education* 35 (1994): 513–37.

Thompson, Janice. *Science of Nutrition*. Boston: Benjamin Cummings, 2010.

Tinto, Vincent. "Stages of Student Departure: Reflections on the Longitudinal Character of Student Leaving." *Journal of Higher Education* 59 (1989): 438–55.

———. *Leaving College: Rethinking the Causes and Cures of Student Attrition, 2nd ed*. Chicago: University of Chicago Press, 1993.

Tobin, Kenneth, Deborah Tippins, and Alejandro Gallard. *Research on Instructional Strategies for Teaching Science. Handbook of Research on Science Teaching and Learning*. New York: MacMillan, 1994.

Treeger, Madge Lawrence, and Karen Levin Coburn. *Letting Go*. New York: Harper Perennial, 1997.

Trefil, James, and Robert Hazen. *Science Matters: Achieving Scientific Literacy*. Harpswell, ME: Anchor Publishers, 2009.

Tyser, R., and W. Cerbin. "Critical Thinking Exercises for Introductory Biology Courses." *Bioscience* 41 (1991): 41–46.

VanDeMark, Brian. *Pandora's Keepers: Nine Men and the Atomic Bomb*. Boston: Back Bay Books, 2005.

Van Gennep, Arnold. *The Rites of Passage*. Chicago: University of Chicago Press, 1960.

Velikovsky, Immanel. *Worlds in Collision*. New York: Pocket Books, 1950.

Vincent, Julian. *Structural Biomaterials*. Princeton, NJ: Princeton University Press, 1990.

Walton, Douglas. *Argument Structure: A Pragmatic Theory*. Toronto, Canada: University of Toronto Press, 1996.

Whitam, F. L., M. Diamond, and J. Martin. "Homosexual Orientation in Twins: A Report on 61 Pairs and Three Triplet Sets." *Archives of Sexual Behavior* 22, no. 3 (1993): 187–206.

Winkle-Wagner, Rachelle. "The Perpetual Homelessness of College Experiences: Tensions between Home and Campus for African-American Women." *Review of Higher Education* 33 (2009): 1–36.

Wolfe, J., "Institutional and External Influences on Social Integration in the Freshman Year. *Journal of Higher Education* 62 (1991): 412–36.

Zahn, C., and M. Miller. "Excess Length of Stay, Charges, and Mortality Attributable to Medical Injuries as a Result of Hospitalization." *American Medical Association* 14 (2005): 1868–74.

Zakaria, F. "Are America's Best Days Behind Us?" *Time*, March 31, 2011. http://fareedzakaria.com/2011/03/03/are-americas-best-days-behind-us/.

Zeldin, Theodore. *An Intimate History of Humanity*. New York: HarperCollins, 1994.

Index

About the Author

Peter A. Daempfle, PhD, has taught biology, anatomy and physiology, human genetics, science issues, and science education courses for over twenty years. Dr. Daempfle has held faculty positions at Hobart and William Smith Colleges and Western New England University, and is currently an associate professor of biology at the State University of New York, College of Technology, at Delhi. He earned his PhD in science education and his MS in biology at the University at Albany, State University of New York; MS in Education from the College of Saint Rose; and BA in Biology from Hartwick College. He was class valedictorian of both Forest Hills High School, Queens, New York, in 1988 and Hartwick College, Oneonta, New York, in 1992 and graduated summa cum laude with departmental distinction in biology and German. Born in the Ridgewood section of Brooklyn, New York, in 1970, Dr. Daempfle was a child of German immigrants.

Dr. Daempfle was the first science education researcher to use qualitative and quantitative approaches to study the academic transition between secondary and postsecondary biology programs. His journal articles are cited extensively and used in contemporary studies throughout science education literature. A scholar in his field, Dr. Daempfle has authored several journal articles; various science reviews; a laboratory manual books; and lectures to scientific and general audiences.

He is the author of the book *Science and Society: Scientific Thought and Education for the 21st Century*; a textbook, *Essential Biology for Allied Health*, and *Anatomy and Physiology: A Guide to the Human Body*, a laboratory manual. He has authored numerous publications in science and science education, focusing on the development of scientific reasoning and retention of students in science. His current research focuses on the tenuous transition between secondary and postsecondary science programs, and human biology and microbiological applications.

From 2001 to 2009 he was a science advisor and consultant to develop standards and assessments required by the Bush administration's No Child Left Behind Act (NCLB). Dr. Daempfle focused on science content applications to psychometrics and test design in relation to standards development on his educational research side. This new text contributes to the efforts of the standards-based reform movement to improve national science literacy and to advance the importance of scientific thinking.